Understanding the Normal Distribution

Christian A. Hume

ISBN 978-976-8223-36-4

Contents

d. Test the ordered pair that you identified in part *c* in the inequality you wrote in part *b*. Does the point satisfy the inequality?

e. Decide whether (1) your current bridge design will allow 13-foot-tall vehicles to pass on the road beneath it or if (2) your bridge must be redesigned. Explain your reasoning in either case.

If you decided on choice (1), move on to part *f*. If you decided on choice (2), skip to part *g*.

f. If you decided that your current bridge design is acceptable, you will need to post the maximum clearance on a sign on the bridge. Find the maximum height of a vehicle that can freely pass beneath the bridge.

g. If you decided that your bridge must be re-designed, then do so. You should retain the arch shape as half an ellipse, and the base of the arch should remain 40 feet wide. Due to restrictions on the bed of the road that crosses over the bridge, the highest point of the bridge's arch may be no more than 18 feet above the center line of the two-lane road below. Give the equation of the ellipse that defines your re-designed arch.

Preface

During the course of my teaching at the Department of Economics at the University of the West Indies Saint Augustine Campus, I observed numerous students routinely getting into statistical difficulties because of an insufficient command of the fundamentals of the normal distribution. I was thereby moved to put something into writing on the normal distribution, and my own course notes eventually morphed into this book.

Nothing is taken for granted here, so we start from the very basics – explaining the concepts of the mean and the variance/standard deviation in chapter 1. The derivation of the standard normal distribution is then explained in chapter 2, setting the stage for chapter 3 where the use of all three positive versions of the z-table (to two decimal places) is outlined for finding probabilities as well as for deriving the value of a random variable when the probability associated with the value is known.

Chapter 4 deals with linear combinations of independent normal variables, chapter 5 outlines the sampling distributions, leading straight into the subject matter of chapters 6 and 7 where the sampling distributions are used in the construction of confidence intervals and in hypothesis tests.

Confidence Intervals is the subject of chapter 6 and Hypothesis Tests the subject of chapter 7. In this book, a clear and deliberate effort is made to present the hypothesis test as the flip-side of the confidence interval. The one-sided confidence interval, although not frequently taught or required for exam purposes, is presented along with its flip-side – the one-tailed hypothesis test. The typical two-sided confidence interval is presented along with its flip-side – the two-tailed hypothesis test.

Following on from this, I present in this book what I call *The Sampling Distribution of the Null Hypothesis*. This concept expands on the idea of explicitly linking the hypothesis test to the confidence interval by deriving the real or implied sampling distribution which would have been used to construct the corresponding confidence interval for the hypothesis test under consideration. The intention is to elucidate the test statistic and its derivation in the conduct of the hypothesis test, eliminating the need to cram intricate formulae. The critical value is also illuminated by this approach.

It is my sincere desire and honest expectation that this book will make your understanding of the normal distribution so clear that you will be able to use it with conviction wherever you may be required to do so, and competently carry out other work whose execution requires a firm grasp of the principles outlined herein. Best wishes on your journey.

Christian A. Hume
(August 2011)

Dedicated to the memory of my high school Statistics teacher at Presentation College San Fernando - Dr. Gunness Ramdath, and to the many people who erroneously believe that they cannot do statistics.

1
Parameters of the Normal Distribution

The normal distribution is the most important and frequently used probability distribution in statistics. It is referred to as 'normal' because it occurs very frequently in the natural world. In other words, it is 'normal' for repeated or multiple measurements of a characteristic or condition to be distributed in such a way as to fit the mathematical model which forms the basis of this particular distribution. However, before any distribution can properly be described as 'normal', it must possess two very important characteristics. It must have a **mean** as well as a **variance** or **standard deviation**. The mean and the variance/standard deviation are not by any means exclusive to the normal distribution, and the mere possession of these characteristics doesn't necessarily make a distribution 'normal', there being other very necessary attributes that define 'normality'. Think of the wings on an airplane. The mere possession of wings doesn't make you an airplane (you may be a bird or a butterfly), but to be an airplane, you MUST have wings (and engines, and wheels, and everything else that makes an airplane). So think of the mean and the variance/standard deviation as the 'wings' of the normal distribution. In the jargon of statistics, we say that the mean and the variance/standard deviation are the **parameters** of the normal distribution.

The Mean

The mean of a particular data set is what we commonly refer to in everyday language as the 'average'. It is found by summing the values of ALL the elements of the data set under consideration, and then dividing the total by the number of elements in the data set. Consider the following example:

Example 1.1
Find the mean of the following data set: 2, 4, 6, 8, 10

The first step is to SUM all of the elements: $2 + 4 + 6 + 8 + 10 = 30$

We then divide this total by the number of elements in the data set. We can see that there are five elements in this particular data set, so we divide 30 by 5 to obtain the mean:
$\frac{30}{5} = 6$. Therefore our mean is 6

Let's go through another example:

Example 1.2
Find the mean of the following data set: 123.45, 114.32, 128.50, 101.99

Again, we start by summing the elements themselves:
123.45 + 114.32 + 128.50 + 101.99 = 468.26

As before, we divide the total by the number of elements in the data set. There are four elements in this particular data set, so we divide 468.26 by 4 to obtain the mean:

$$\frac{468.26}{4} = 117.07$$

Our mean is therefore 117.07

Another example:

Example 1.3
Find the mean of the following data set: 0.54, 0.823, 0.36, 0.11, 0.095, 0.44

Summing the elements, we have: 0.54 + 0.823 + 0.36 + 0.11 + 0.095 + 0.44 = 2.368

Dividing the total by the number of elements: $\frac{2.368}{6} = 0.3947$

Our mean here is therefore 0.3947

What about negative numbers?

Example 1.4
Find the mean of the following data set: - 3, - 5, - 9

Summing the elements: (- 3) + (- 5) + (- 9) = - 3 - 5 - 9 = -17

Dividing the total by the number of elements: $\frac{-17}{3} = $ - 5.66

In this case, the mean is - 5.66

It matters not how many elements there are, nor whether the elements are large, small, decimal, fraction, negative, or positive. The procedure is the same - sum the elements, then divide the total by the number of elements.

Exercise 1.1
Find the mean of each of the following data sets:

(1) 3, 4, 7, 8

(2) 2.35, 7.09, 13.24, 16.21

(3) 0.01, 0.035, 0.009, 0.14

(4) -5, -7, -12, -14, -12

(5) 1,234; 996; 837; 439; 667; 901

(6) 27.48, 56.35, 68.79, 34.16, 66.23

(7) - 0.35, - 0.7, - 0.83

(8) 10,283; 8,999; 4,056; 3,223

(9) 4.04, -2.25, 3.57, -6.056

(10) 16, 43, 27, 32, 65, 11, 38, 59, 61, 77, 84

The Variance/Standard Deviation

Mathematically speaking, the variance and the standard deviation represent a classic case of 'six of one, half a dozen of the other'. There really is no practical difference between the two. Normally, only one is used at a time, but one is simply the square of the other, or the square root of the other, depending on which direction you are looking from.

The Standard Deviation is the square root of the Variance
The Variance is the square of the Standard Deviation

Be very careful to note which is the 'square', and which is the 'square root'. If we have the variance, we derive the standard deviation simply by finding its square root. When we have the standard deviation, we find the variance by finding its square. Thus in mathematical symbols:

$$Standard\ Deviation\ =\ \sqrt{Variance}$$
$$Variance\ =\ (Standard\ Deviation)^2$$

So, what exactly is this standard deviation? We will illuminate the concept by examining three data sets with the same mean. Consider the following:

Data Set 1: 50, 50, 50
Data Set 2: 49, 50, 51
Data Set 3: 40, 50, 60

Let us find the mean of each of these three data sets:

Data Set 1: Mean = $\dfrac{50+50+50}{3} = \dfrac{150}{3} = 50$

Data Set 2: Mean = $\dfrac{49+50+51}{3} = \dfrac{150}{3} = 50$

Data Set 3: Mean = $\dfrac{40+50+60}{3} = \dfrac{150}{3} = 50$

It is clear that all three data sets have an identical mean of 50. But the data sets are not the same. So, what is different about the data sets? Well, the first thing we notice is that in data set 1, all the elements are the same. Each element is 50. Therefore, we can say that there is *no variation* among the elements. We can also say that the *variance* among elements is equal to zero. Since the mean is also equal to 50, we can say that for each element, there is no *deviation* from the mean.

In data set 2, we have slightly more *variation* among the elements, whose values *vary* from 49 to 51, and for each of the elements 49, 50, and 51, we also some *deviation* from the mean of 50.

In data set 3, we have even more *variation* among the elements, whose values *vary* from 40 straight up to 60. For each of the elements 40, 50, and 60, there is even more *deviation* from the mean of 50 than there was in data set 1.

We can say that there is no dispersion in data set 1, since all the elements have the same value. Set 2 possesses slightly more dispersion, the values 49 and 51 being respectively less than and greater than the mean 50 by one. Set 3 is even more widely dispersed, with the values 40 and 60 being respectively less than and greater than the mean 50 by ten.

So, although the three data sets have the same mean of 50, they each possess different degrees of variation among the elements, they each possess different degrees of deviation between each element and the mean, and they are each differently dispersed.

The concept of the variance/standard deviation is one of dispersion, or spread. Two data sets can have the same mean, but one data set may have elements that are more widely dispersed than the elements in the other data set.

In statistics, the process of evaluating the variance of a data set is essentially that of putting a precise numerical value on the degree of *variance, spread, or dispersion* of a data set. It is the process of putting a numerical value on the degree to which a random value in the data set *deviates* from the mean.

Conceptually, the standard deviation and the variance are connected because for any data set, the greater the deviation of individual values from the mean, the greater the overall variance among elements would be, and the greater the overall variance between the highest and lowest elements would be. Mathematically, they are connected by the square/square root relationship we observed earlier on page 3.

Mathematical Calculation of Variance/Standard Deviation

To find the precise mathematical value of the standard deviation, there are some steps that we must follow:

(a) Find the mean
(b) Find the deviation of each element in the data set from the mean
(c) Square each deviation
(d) Sum the squared deviations
(e) Divide by the number of elements
(f) Find the square root

Proceeding through steps *(a)* to *(e)* would give us the variance, and the final step *(f)* would give the standard deviation, with the variance already in place up to step *(e)*. We now illustrate the procedure by finding the variance and the standard deviation of each of the three data sets under consideration.

Data Set 1

(a) Find the mean
From earlier work, we know that the mean is 50

(b) Find the deviation of each element in the data set from the mean

x	$\bar{x}$	$x - \bar{x}$
50	50	0
50	50	0
50	50	0

For each element (x), the deviation from the mean ($\bar{x}$) is zero.

(c) Square each deviation

x	$\bar{x}$	$x - \bar{x}$	$(x - \bar{x})^2$
50	50	0	0
50	50	0	0
50	50	0	0

(d) Sum the squared deviations

x	$\bar{x}$	$x - \bar{x}$	$(x - \bar{x})^2$
50	50	0	0
50	50	0	0
50	50	0	0

$$\sum 0$$

(e) Divide by the number of elements

$$\frac{0}{3} = 0.$$

At this point, we have the variance. We find the standard deviation by proceeding to the next step.

(f) Find the square root

$$\sqrt{0} = 0.$$

So, we have a variance of 0, and a standard deviation of 0. These values resonate perfectly with our previous analysis of the concepts of variance and deviation as applied to this particular data set where all the elements are the same.

The only situation that would give us a variance or a standard deviation of 0 is one where all the elements are the same.

Data Set 2

(a) *Find the mean*
     From earlier work, we know that the mean is 50

(b) *Find the deviation of each element in the data set from the mean*

x	$\bar{x}$	$x - \bar{x}$
49	50	-1
50	50	0
51	50	1

(c) *Square each deviation*

x	$\bar{x}$	$x - \bar{x}$	$(x - \bar{x})^2$
49	50	-1	1
50	50	0	0
51	50	1	1

(d) *Sum the squared deviations*

x	$\bar{x}$	$x - \bar{x}$	$(x - \bar{x})^2$
49	50	-1	1
50	50	0	0
51	50	1	1

$$\sum \ 2$$

(e) Divide by the number of elements

$$\frac{2}{3} = 0.6667$$

At this point, we have the variance. We find the standard deviation by proceeding to the next step.

(f) Find the square root
$$\sqrt{0.6667} = 0.8167$$

So, we have a variance of 0.6667, and a standard deviation of 0.8167.

Data Set 3

(a) Find the mean
From earlier work, we know that the mean is 50

(b) Find the deviation of each element in the data set from the mean

x	$\bar{x}$	$x - \bar{x}$
40	50	-10
50	50	0
60	50	10

(c) Square each deviation

x	$\bar{x}$	$x - \bar{x}$	$(x - \bar{x})^2$
40	50	-10	100
50	50	0	0
60	50	10	100

(d) Sum the squared deviations

x	$\bar{x}$	$x - \bar{x}$	$(x - \bar{x})^2$
40	50	-10	100
50	50	0	0
60	50	10	100

$$\sum 200$$

(e) Divide by the number of elements
$$\frac{200}{3} = 66.67$$

At this point, we have the variance. We find the standard deviation by proceeding to the next step.

(f) Find the square root
$$\sqrt{66.67} = 8.17$$

So, we have a variance of 66.67, and a standard deviation of 8.17.

The final values for all three data sets concur with the situational analyses earlier, where we saw that data set 1 had a no variation, (hence variance = 0), and for each element, there was no deviation from the mean (standard deviation = 0).

The variance and standard deviation of data set 3 were greater than those of data set 2, conforming to our prior reasoning that data set 3 was more dispersed than data set 2. Here we have just put precise mathematical values on the degree of dispersion of each data set, and the degree of deviation from the mean of the elements in each set.

Let us now go through an example with more elements:

Example 1.5
Find the standard deviation of the following data set:
2.43, 4.32, 1.66, 6.98, 3.81, 5.55, 7.63, 2.39

We follow the steps as outlined previously:

(a) Find the mean
$$\bar{x} = \frac{2.43 + 4.32 + 1.66 + 6.98 + 3.81 + 5.55 + 7.63 + 2.39}{8} = \frac{34.77}{8} = 4.35$$

(b) Find the deviation of each element in the data set from the mean

x	$\bar{x}$	$x - \bar{x}$
2.43	4.35	- 1.92
4.32	4.35	- 0.03
1.66	4.35	- 2.69
6.98	4.35	2.63
3.81	4.35	- 0.54
5.55	4.35	1.20
7.63	4.35	3.28
2.39	4.35	- 1.96

(c) Square each deviation

x	$\bar{x}$	$x - \bar{x}$	$(x - \bar{x})^2$
2.43	4.35	- 1.92	3.6864
4.32	4.35	- 0.03	0.0009
1.66	4.35	- 2.69	7.2361
6.98	4.35	2.63	6.9169
3.81	4.35	- 0.54	0.2916
5.55	4.35	1.20	1.4400
7.63	4.35	3.28	10.7584
2.39	4.35	- 1.96	3.8416

(d) Sum the squared deviations

x	$\bar{x}$	$x - \bar{x}$	$(x - \bar{x})^2$
2.43	4.35	- 1.92	3.6864
4.32	4.35	- 0.03	0.0009
1.66	4.35	- 2.69	7.2361
6.98	4.35	2.63	6.9169
3.81	4.35	- 0.54	0.2916
5.55	4.35	1.20	1.4400
7.63	4.35	3.28	10.7584
2.39	4.35	- 1.96	3.8416
		$\sum$	34.1719

(e) Divide by the number of elements

$$\frac{34.1719}{8} = 4.27$$

(f) Find the square root

$$\sqrt{4.27} = 2.07$$

The variance is therefore 4.27, the standard deviation being 2.07.

Let us now find the standard deviation of a set of *negative* numbers:

Example 1.6
Find the standard deviation of the following data set: - 3, - 5, - 7

(a) Find the mean

$$\bar{x} = \frac{-3 + (-5) + (-7)}{3} = \frac{-3 - 5 - 7}{3} = \frac{-15}{3} = -5$$

(b) Find the deviation of each element in the data set from the mean

x	$\bar{x}$	$x - \bar{x}$
- 3	- 5	2.00
- 5	- 5	0.00
- 7	- 5	- 2.00

(c) Square each deviation

x	$\bar{x}$	$x - \bar{x}$	$(x - \bar{x})^2$
- 3	- 5	2.00	4.00
- 5	- 5	0.00	0.00
- 7	- 5	2.00	4.00

(d) Sum the squared deviations

x	$\bar{x}$	$x - \bar{x}$	$(x - \bar{x})^2$
- 3	- 5	2.00	4.00
- 5	- 5	0.00	0.00
- 7	- 5	2.00	4.00

$$\sum 8.00$$

(e) Divide by the number of elements

$$\frac{8.00}{3} = 2.67$$

At this point we have the variance. We now find the standard deviation by proceeding to the final step.

(f) Find the square root

$$\sqrt{2.67} = 1.63$$

So, the variance is 3.1156, and the standard deviation is 1.63. Note that both the variance and the standard deviation are *positive,* even though the data set was composed entirely of negative numbers.

*The variance and the standard deviation of any data set are **ALWAYS** positive*

Exercise 1.2
Find the standard deviation of each of the following data sets: (these are the identical data sets used in Exercise 1.1)

(1) 3, 4, 7, 8

(2) 2.35, 7.09, 13.24, 16.21

(3) 0.01, 0.035, 0.009, 0.14

(4) -5, -7, -12, -14, -12

(5) 1,234; 996; 837; 439; 667; 901

(6) 27.48, 56.35, 68.79, 34.16, 66.23

(7) - 0.35, - 0.7, - 0.83

(8) 10,283; 8,999; 4,056; 3,223

(9) 4.04, -2.25, 3.57, -6.056

(10) 16, 43, 27, 32, 65, 11, 38, 59, 61, 77, 84

Notation of the Normal Distribution

In the written language of the Normal Distribution, the mean is represented by the Greek symbol μ (mu), while the standard deviation is represented by the Greek symbol σ (sigma). The variance is therefore represented by the symbol σ^2 (sigma squared). Any Normal Distribution can be specified either by stating its mean and standard deviation, or its mean and variance. In this book we will use the latter convention since it greatly simplifies matters when combining Normal Distributions (chapter 4). The mean of the Normal Distribution is also referred to as its **expected value**, and so the mean of a general normal random variable X is written as E(X). Similarly, the variance of this general normal random variable X is written as Var(X).

The general random variable X which is normally distributed with mean μ, and standard deviation σ (hence variance σ^2), is written as:

$$X \sim N(\mu, \sigma^2)$$

For a Normal Distribution with mean μ and variance σ^2, E(X) = μ, and Var(X) = σ^2

The Shape and Characteristics of the Normal Distribution

In the normal distribution, the data values are arranged in ascending order, starting from the smallest value, right up to the largest value. Each value has an associated frequency (the number of times we encounter that value in the data set). The mean lies at the centre of all data values. It is also the data value that occurs with the highest frequency (it occurs most often). As you move away from the mean in both directions, the frequencies of the data values decrease at the same rate on both sides of the mean. These features give the normal distribution its characteristic symmetric 'bell shape'. The diagram on the following page presents the shape and characteristics of a generic normal distribution with mean μ and standard deviation σ (variance σ^2).

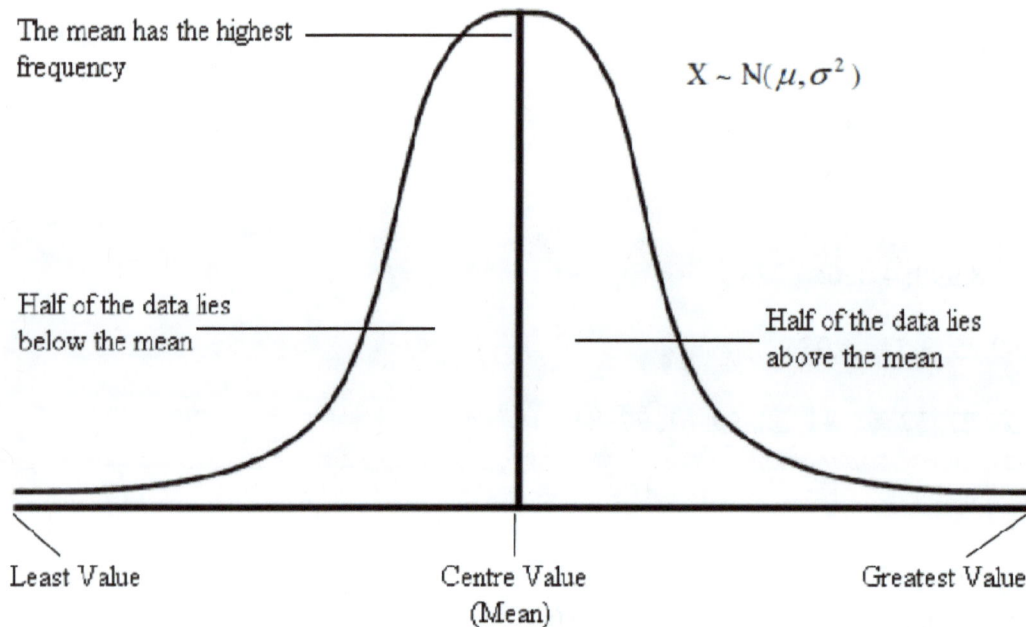

The mean has the highest frequency

$X \sim N(\mu, \sigma^2)$

Half of the data lies below the mean

Half of the data lies above the mean

Least Value   Centre Value (Mean)   Greatest Value

Of course, it would be very rare in the real world to find a set of data whose elements conform *exactly* to *every* characteristic of the Normal Distribution. But very often, the data conforms closely enough to the these chracterics to allow us to use the Normal Distribution as a **model** for a particular data set. We are then able to utilize the mathematical properties of this model to estimate probabilites and values, which can then be used as tools for decision-making.

Exercise 1.3

Express the following normally distributed random variables using the statistical notation of the Normal Distribution:

(1) A random variable X is normally distributed with a mean of 23 and a variance of 14.

(2) A random variable W is normally distributed with a mean of 0.003 and a standard deviation of 0.12.

(3) A random variable P is normally distributed with a mean of 1,256 and a variance of 576.

(4) A random variable Q is normally distributed with a mean of 7,653,256 and a standard deviation of 1,245.

(5) A random variable Y is normally distributed with a mean of 25.43 and a variance of 16.

(6) A random variable B is normally distributed with a mean of -13.38 and a standard deviation of 3.68.

(7) A random variable F is normally distributed with a mean of 2.337 and a variance of 4.41.

(8) A random variable K is normally distributed with a mean of 24,643 and a standard deviation of 3,005.

(9) A random variable T is normally distributed with a mean of - 0.56 and a variance of 0.99.

(10) A random variable V is normally distributed with a mean of -134 and a standard deviation of 12.

2
The Standard Normal Distribution

Why we need a Standard Normal Distribution

The mathematics by which the probabilities for the normal distribution are calculated is extremely tedious – tedious enough that we will leave it well outside the scope of this book. Additionally, there is no limit to the possible values of the defining parameters of the normal distribution – the mean and the variance/standard deviation. A randomly imagined normal distribution can have a mean of any value, and a variance/standard deviation of any value. Here are some possible normal distributions:

$X \sim N(150, 15^2)$
$X \sim N(12.56, 4.354^2)$
$X \sim N(0.00456, 0.00034^2)$
$X \sim N(-3.567, 1.32^2)$
$X \sim N(1,543,341, 122,876^2)$

In fact, any two numbers that you can think of, regardless of their relative magnitude, can take the values of the parameters of a normal distribution - the only limitation being that the variance/standard deviation cannot be negative. The mean however, may be negative. The infinite number of possible combinations of the values of the parameters of the normal distribution, coupled with the extremely tedious mathematics involved in the probability calculations for each pair of values meant that a more straightforward route had to be found to manipulate the probabilities of the normal distribution. Fortunately for us all, the very nature of the distribution itself makes this possible.

This is where the standard normal distribution enters the picture. The essential idea is that certain special conditions that are common to **ALL** normal distributions would be used to create a standardized normal distribution that can be used to model **any** normal distribution.

Standardizing any value on any Normal Distribution

For any normal distribution, we can express any numerical value associated with that distribution in terms of how far it lies from the mean. We can do this by expressing how many standard deviations away from the mean the value lies. Suppose we have two normal distributions as follows:

$X \sim N(100,10^2)$
$Y \sim N(100,15^2)$

We want to analyze the value 110 that lies on both distributions. On both distributions, the value 110 lies 10 units away from the mean (110 − 100 = 10). However, in terms of the standard deviation, the situation becomes a bit more interesting.

On the first distribution, the value 110 lies 1 standard deviation away from the mean $\left(\dfrac{110-100}{10} = \dfrac{10}{10} = 1 \right)$.

On the second distribution, the value 110 lies 0.67 standard deviations away from the mean $\left(\dfrac{110-100}{15} = \dfrac{10}{15} = 0.67 \right)$.

The mathematics underlying the normal distribution would also tell us how much of the data values lies within 0.67 standard deviations of the mean. In other words, it would tell us the probability that a randomly selected data value would lie within 0.67 standard deviations from the mean.

Suppose that on the same two normal distributions, we wish to analyze the value 90. On both distributions, the value 90 lies -10 units away from the mean (90 − 100 = - 10). However, in terms of the standard deviation, we have the following situations:

On the first distribution, the value 90 lies - 1 standard deviations away from the mean $\left(\dfrac{90-100}{10} = \dfrac{-10}{10} = -1 \right)$.

On the second distribution, the value 90 lies -0.67 standard deviations away from the mean $\left(\dfrac{90-100}{15} = \dfrac{-10}{15} = -0.67 \right)$.

We can see that the value 90 is less than the mean 100. So that in terms of how far the value 90 lies from the mean, the 'negative' sign tells us the value under consideration is less than the mean. Therefore on the first distribution, the value 110 lies one standard deviation above the mean, while the value 90 lies one standard deviation below the mean. Similarly, on the second distribution, the value 110 lies 0.67 standard deviations above the mean, while the value 90 lies 0.67 standard deviations below the mean.

Exercise 2.1
If a random variable is normally distributed with a mean of 65 and a standard deviation of 8, how many standard deviations away from the mean do the following values lie?

(1) 67	**(6)** 75
(2) 59	**(7)** 50
(3) 65	**(8)** 47
(4) 66.3	**(9)** 83
(5) 61.25	**(10)** 40

The number of standard deviations away from the mean that a numerical value lies is called its **standardized z-value**. If the numerical value is less than the mean, its standardized z-value will be negative. If the numerical value is greater than the mean, its standardized z-value will be positive.

So in the example where $X \sim N(100,10^2)$, the value 110 has a standardized z-value of 1, while the value 90 has a standardized z-value of - 1. Likewise, in the case where $Y \sim N(100,15^2)$, the value 110 has a standardized z-value of 0.67, while the value 90 has a standardized z-value of - 0.67.

A value on a normal distribution with a standardized z-value of 2.03 therefore lies 2.03 standard deviations *above* the mean. Likewise, a value on a normal distribution with a standardized z-value of - 2.74 lies 2.74 standard deviations *below* the mean.

The process of standardizing any value on a normal distribution is essentially the process of determining how far away from the mean that value lies – how many standard deviations away from the mean the value lies.

Exercise 2.2
How far away from the mean do values on a normal distribution with the following standardized z-values lie? Specify whether the value is above or below the mean.

(1) 0.54	**(6)** -3.01
(2) -1.21	**(7)** 2.31
(3) -0.99	**(8)** -0.56
(4) 1.92	**(9)** 1.13
(5) 2.42	**(10)** -2.07

De-standardizing any standardized z-value

If a normal distribution has a mean of 100 and a standard deviation of 10, and a value on this distribution lies 1 standard deviation above the mean, what is that value? This unknown value has a standardized z-value of 1.

The normal distribution can be represented as: $X \sim N(100,10^2)$. We are essentially trying to determine the value on this distribution that is 1 standard deviation above the mean of 100. Let us represent this unknown value as X_1.

Therefore: $\dfrac{X_1 - 100}{10} = 1$

Cross-multiplying, $X_1 - 100 = 10(1)$

$$X_1 - 100 = 10$$

Carrying the '100' across the equal sign:

$$X_1 = 10 + 100$$
$$= 110$$

Therefore the value on the normal distribution $X \sim N(100,10^2)$ which lies 1 standard deviation above the mean of 100 is 110.

Similarly, if an unknown value on the normal distribution $Y \sim N(100,15^2)$ lies 0.67 standard deviations below the mean, what is that value? This unknown value has a standardized z-value of - 0.67. What we are really doing is determining the value on this distribution which lies 0.67 standard deviations below the mean of 100. Let us represent this unknown value as Y_{-1}.

Therefore: $\dfrac{Y_{-1} - 100}{15} = -0.67$

Cross-multiplying, $Y_{-1} - 100 = 15(-0.67)$

$$Y_{-1} - 100 = -10$$

Carrying the '100' across the equal sign:

$$Y_{-1} = -10 + 100$$
$$= 100 - 10$$
$$= 90$$

Therefore the value on the normal distribution $Y \sim N(100,15^2)$ which lies - 0.67 standard deviations below the mean of 100 is 90.

The standardized z-value of an unknown value on any normal distribution tells us how far away from the mean that value lies – how many standard deviations above or below the mean the value lies. An unknown value with a positive standardized z-value lies z standard deviations above the mean, while an unknown value with a negative standardized z-value lies z standard deviations below the mean

Exercise 2.3

If a random variable is normally distributed with a mean of 65 and a standard deviation of 8, determine the values that lie:

(1) 1.06 standard deviations above the mean
(2) 2.33 standard deviations above the mean
(3) 0 standard deviations away from the mean
(4) 0.66 standard deviations below the mean
(5) 1.65 standard deviations below the mean
(6) 2.99 standard deviations below the mean
(7) 2.99 standard deviations above the mean
(8) 1.34 standard deviations above the mean
(9) 3.03 standard deviations below the mean
(10) 0.79 standard deviations above the mean

The Standard Normal Distribution: z ~ N(0,1)

The standard normal distribution is defined as z ~ N(0,1). The z-value on this distribution tells us how many standard deviations away from the mean the value lies. Since the location of a z-value is defined with reference to the mean, then the mean itself would be located l0 standard deviations away from the mean. This is the reason why the mean of the standard normal distribution is 0. The standard deviation is equal to 1 because 1 is the only value that would allow the distance of a z-value away from the mean in units to be equal to the distance of the z-value from the mean in standard deviations. For example, if a normal distribution has a mean of 0 and a standard deviation of 2, then the value 1 on this distribution would be 0.5 standard deviations above from the mean $\left(\frac{1-0}{2} = \frac{1}{2} = 0.5\right)$. If the mean is '0' and the standard deviation is 1, then the value 1 would be 1 standard deviation above the mean $\left(\frac{1-0}{1} = \frac{1}{1} = 1\right)$. In both cases, the value 1 is 1 unit away from the mean of 0 (1 - 0 = 1). However, the number of standard deviations away from the mean 0 that the value 1 lies depends on the value of the standard deviation itself. If the standard deviation is equal to 1, then the value 1 would lie 1 standard deviation above the mean 0. Similarly, if the standard deviation is 1, then the value -1 would lie 1 standard deviation below the mean 0.

The mean of the standard normal distribution is 0, and the standard deviation is 1. The variance is also 1.

Exercise 2.4
Consider the standard normal distribution z ~ N(0,1). How far away from the mean does each of the following z-values lie? Specify whether the value lies above, below, or on the mean.

(1) 0
(2) 1.16
(3) -0.77
(4) 2.09
(5) -1.87

(6) 4.01
(7) -3.96
(8) 0.23
(9) -1.44
(10) -2.67

The 68-95-99.7 rule
The mathematics underlying the normal distribution tells us that:
  (a) 68% of the data lies within one standard deviation of the mean
  (b) 95% of the data lies within two standard deviations of the mean
  (c) 99.7% of the data lies within three standard deviations of the mean

We know that we can express the values in *any* normal distribution in terms of how far they lie from the mean. This distance from the mean is expressed in terms of the number of standard deviations away from the mean. This is essentially what the z-table and the z-curve are about. So for example, we know that 68% of the data values in any normal distribution lie within 1 standard deviation of the mean. Graphically, this would be represented as:

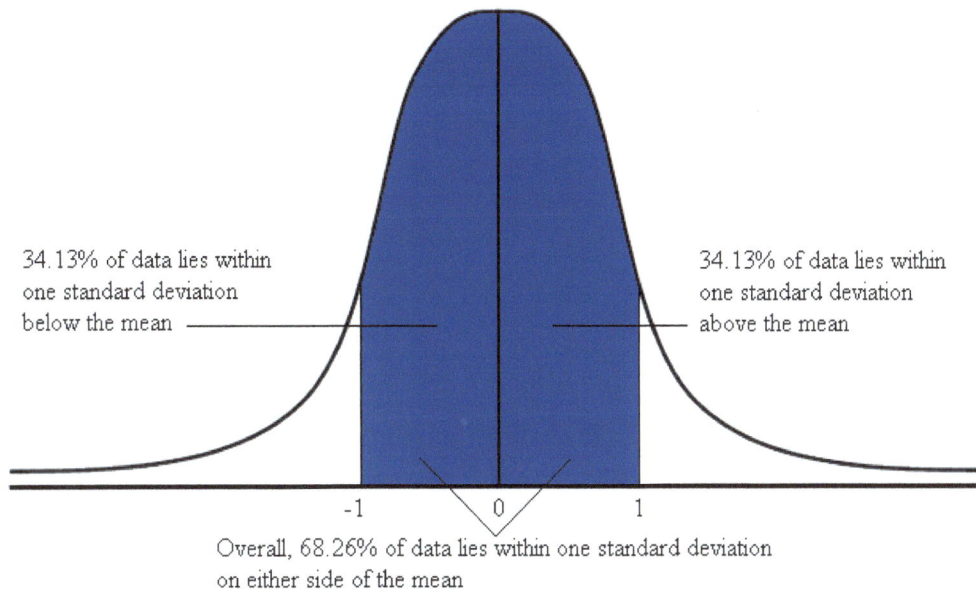

34.13% of data lies within one standard deviation below the mean

34.13% of data lies within one standard deviation above the mean

-1   0   1

Overall, 68.26% of data lies within one standard deviation on either side of the mean

When we say 'within one standard deviation of the mean' we mean one standard deviation on either side of the mean. This means that the region from one standard deviation below the mean (z = -1) to one standard deviation above the mean (z = 1) contains approximately 68% of the data values. The exact value to two decimal places is 68.26%. This 68.26% is divided equally between both sides of the region in blue, with the mean of course as the central dividing line. This means that the region from the mean itself (z = 0) to the value one standard deviation below the mean (z = -1) contains 34.13% of the data values. Similarly, the region from the mean (z = 0) to the value one standard deviation above the mean (z = 1), also contains 34.13% of the data values. The symmetrical nature of the normal distribution means that if 34.13% of the data values lie between the mean (z = 0) and the value one standard deviation below the mean (z = -1), then 34.13% of the data values also lie between the mean (z = 0) and the value that lies one standard deviation above the mean (z = 1).

This same logic can be applied to values two standard deviations above or below the mean:

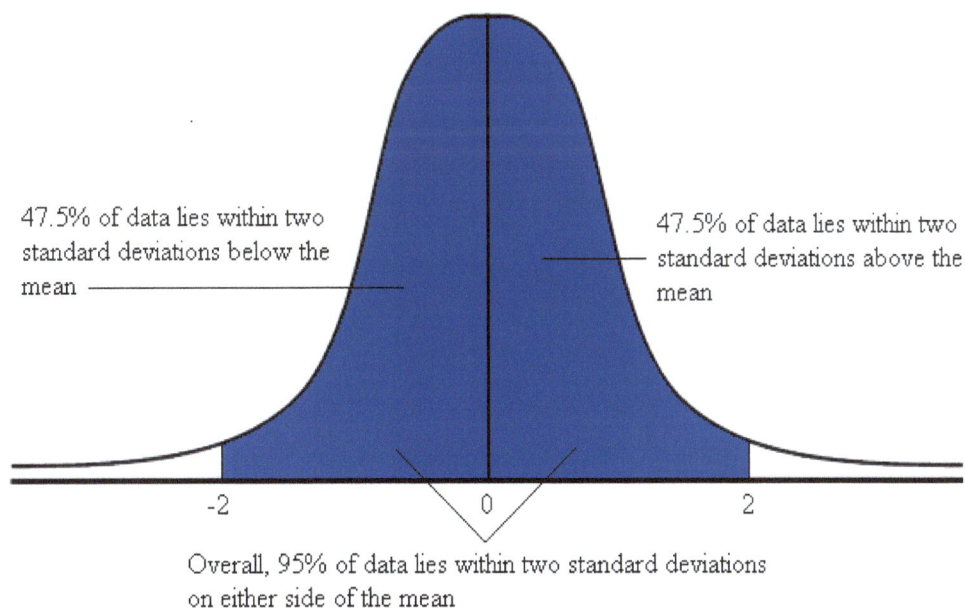

47.5% of data lies within two standard deviations below the mean

47.5% of data lies within two standard deviations above the mean

-2   0   2

Overall, 95% of data lies within two standard deviations on either side of the mean

The region from two standard deviations below the mean ($z = -2$) to two standard deviations above the mean ($z = 2$) contains approximately 95% of the data values. This 95% is divided equally between both sides of the region in blue, with the mean of course as the central dividing line. This means that the region from the mean itself ($z = 0$) to the value 2 standard deviations below the mean ($z = -2$) contains 47.5% of the data values. Similarly, the region from the mean ($z = 0$) to the value one standard deviation above the mean ($z = 2$), also contains 47.5% of the data values.

In the case of values three standard deviations above or below the mean:

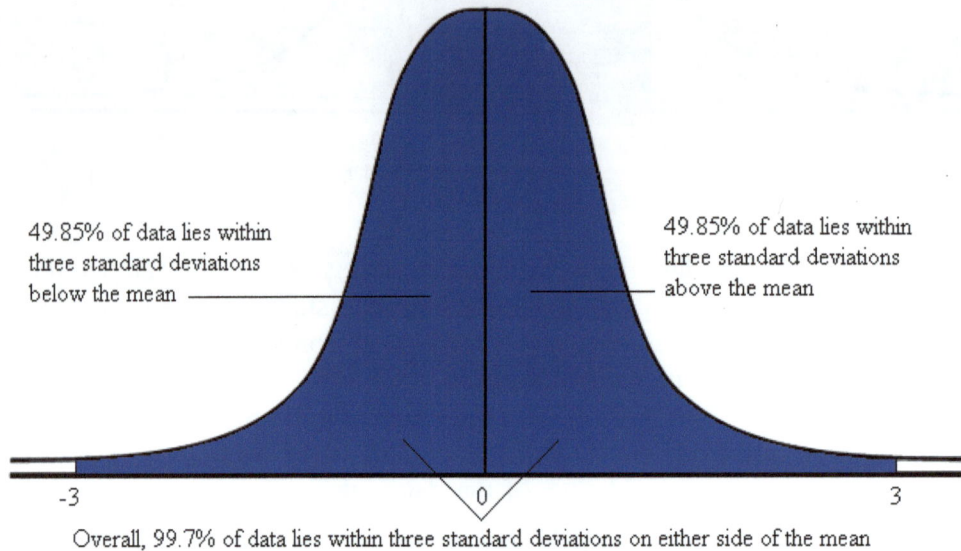

49.85% of data lies within three standard deviations below the mean

49.85% of data lies within three standard deviations above the mean

-3   0   3

Overall, 99.7% of data lies within three standard deviations on either side of the mean

The region from three standard deviations below the mean ($z = -3$) to three standard deviations above the mean ($z = 3$) contains approximately 99.7% of the data values. This 99.7% is divided equally between both sides of the region in blue, with the mean of course as the central dividing line. This means that the region from the mean itself ($z = 0$) to the value three standard deviations below the mean ($z = -3$) contains 49.85% of the data values. Similarly, the region from the mean ($z = 0$) to the value three standard deviation above the mean ($z = 3$), also contains 49.85% of the data values.

This reasoning and procedure can in fact be applied to values any number of standard deviations above or below the mean, and this is the foundation of the z-table. The mathematics that allowed mathematicians to determine that 34.13% of the data values for a normal distribution lie between the mean and a value one standard deviation above or below the mean was used to calculate similar percentages for **all** convenient numbers of standard deviations above or below the mean. This is essentially what the z-table is. So if you go to z-table 1 on page A2 of the appendices, you would see that the value for the probability that a randomly chosen data value lies between $z = 0$ and $z = 1$ is 0.3413. Similarly, the probability that a chosen data value lies between $z = 0$ and $z = 2$ is 0.4750 and by the same token, the probability of a randomly chosen data value lying between $z = 0$ and $z = 3$ is 0.4985.

What then would be the probability that a randomly chosen data value lies between:
(a) $z = -3$ and $z = 0$?
(b) $z = -2$ and $z = 0$?
(c) $z = -1$ and $z = 0$?

Exercise 2.5

(1) If a random variable is normally distributed with a mean of 155 and a standard deviation of 16.6, determine the standardized z-score of each of the following values:
(a) 157   (b) 173   (c) 149   (d) 138

(2) A random variable P is defined normally such that $P \sim N(0.04, 0.023)$. Determine the standardized z-score of the values:
(a) 0.075   (b) 0.016   (c) 0.055   (d) 0.023

(3) If $T \sim N(-3.2, 2^2)$, how far away from the mean do the following values lie?
(a) -2.7   (b) -4.2   (c) -3.2   (d) -1.9   (e) – 0.34

(4) The annual salaries of the members of a millionaires' club are normally distributed with a mean of \$1,564,342 and a standard deviation of \$250,000. What is the standardized z-score of each of the following annual salaries?
(a) \$2,000,000   (b) \$1,705,348   (c) 1,001,543   d) 1,986,321   (e) 2,105,457

(5) The number of points W scored by a particular basketball team is normally distributed such that $W \sim N(89, 5.5^2)$. How far away from the mean do each of the following scores lie?
(a) 77   (b) 101   (c) 93   (d) 89   (e) 82   (f) 86   (g) 91

(6) For the random variable in question 1, determine the de-standardized values of the following z-scores:
(a) 1.23   (b) -2.15   (c) 0.66   (d) 0   (e) -1.02

(7) For the random variable in question 2, determine the de-standardized values of the following z-scores:
(a) -2.07   (b) -0.77   (c) 0   (d) 0.77   (e) 2.07   (f) 1.44

(8) If $X \sim N(-0.035, 0.025^2)$, find the de-standardized values of the following z-scores:
(a) 1.00   (b) -2.00   (c) 0   (d) -1.45   (e) 2.11   (f) -0.55

(9) The number of colonies of a particular bacteria per 100 millilitres of water at a particular beach is normally distributed with a mean of 84 and a standard deviation of 9. Find the actual number of colonies per 100 millilitres represented by the following z-scores:
(a) -1.33   (b) 2.18   (c) -0.78   (d) 0.12

(10) The heights of cotton trees on a plantation are normally distributed with a mean of 181.4 centimetres and a standard deviation of 8.21 centimetres. Find the height represented by each of the following z-scores:
(a) 0.99   (b) 1.23   (c) -1.87   (d) -0.98   (e) 0   (f) 1.11   (g) -2.03

3

The z-table

Finding the probability that corresponds to a particular z-value

The ability to use the z-table expeditiously is one of the fundamental skills necessary in the solution of Normal Distribution problems. Without this ability, even if you understand the situation presented to you, this understanding cannot be translated into a satisfactory solution to your problem.

There are at least six versions of the z-table that are used to solve Normal Distribution problems. Go to Appendix A, and you will see three versions of the z-table presented, each version giving the positive z-score to two decimal places. While you it is possible to get by using only one version, it is best to develop facility in the use of all three versions that we present in this book. As such, the examples and explanations will involve all three versions, while in actual problems, you are free to choose whichever table best suits your needs for each particular problem.

Suppose we want to find p(z > 1). On the z – curve, this area is represented by:

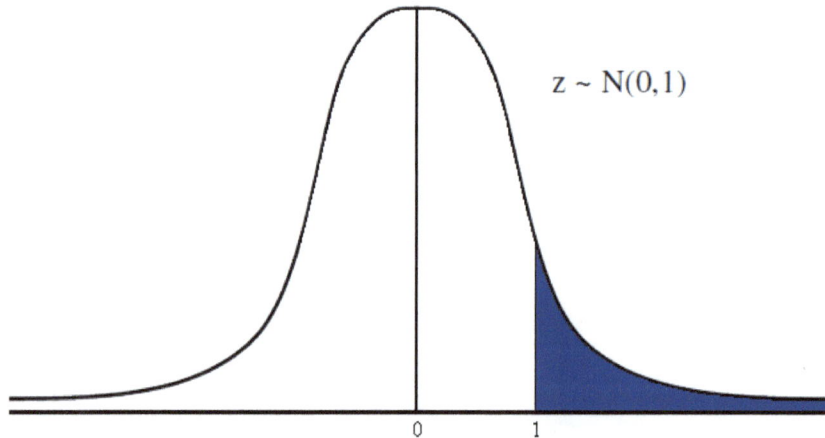

$z \sim N(0,1)$

We can use any of the three versions of the z-table presented to solve this problem.

Using z-table 1

We go to the value $z = 1.00$. We do this by looking for '1.0' in the left hand column, then we look for '.00' in the row at the top of the table. This combination gives us the value '1.00'.

z	.00	.01	.02	.03	.04
0.0	0.0000	0.0040	0.0080	0.0120	0.0160
0.1	0.0398	0.0438	0.0478	0.0517	0.0557
0.2	0.0793	0.0832	0.0871	0.0910	0.0948
0.3	0.1179	0.1217	0.1255	0.1293	0.1331
0.4	0.1554	0.1591	0.1628	0.1664	0.1700
0.5	0.1915	0.1950	0.1985	0.2019	0.2054
0.6	0.2257	0.2291	0.2324	0.2357	0.2389
0.7	0.2580	0.2611	0.2642	0.2673	0.2704
0.8	0.2881	0.2910	0.2939	0.2967	0.2995
0.9	0.3159	0.3186	0.3212	0.3238	0.3264
1.0	0.3413	0.3438	0.3461	0.3485	0.3508
1.1	0.3643	0.3665	0.3686	0.3708	0.3729
1.2	0.3849	0.3869	0.3888	0.3907	0.3925
1.3	0.4032	0.4049	0.4066	0.4082	0.4099
1.4	0.4192	0.4207	0.4222	0.4236	0.4251

Consider the entire area on the right side of the mean 0. This area is equal to 0.5, comprised of the area in light blue between $z = 0$ and $z = 1$, and the area in dark blue to the right of $z = 1$.

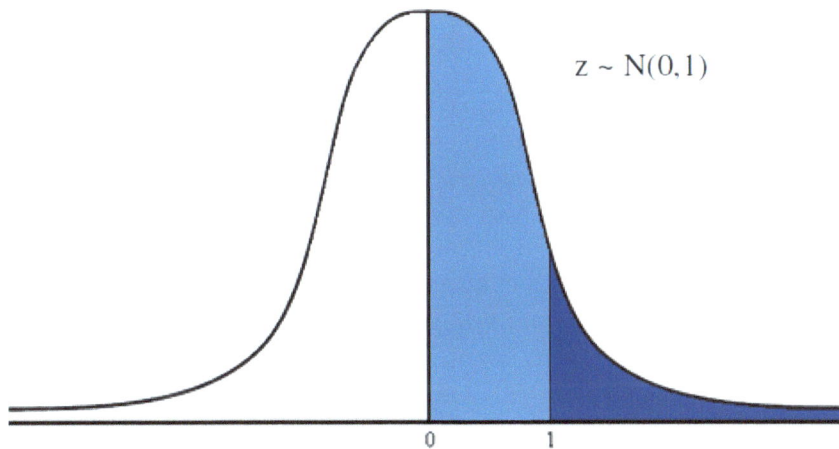

$z \sim N(0,1)$

We can obtain the area in dark blue to the right of $z = 1$ by simply taking away the light blue area between $z = 0$ and $z = 1$ from 0.5, leaving us with:

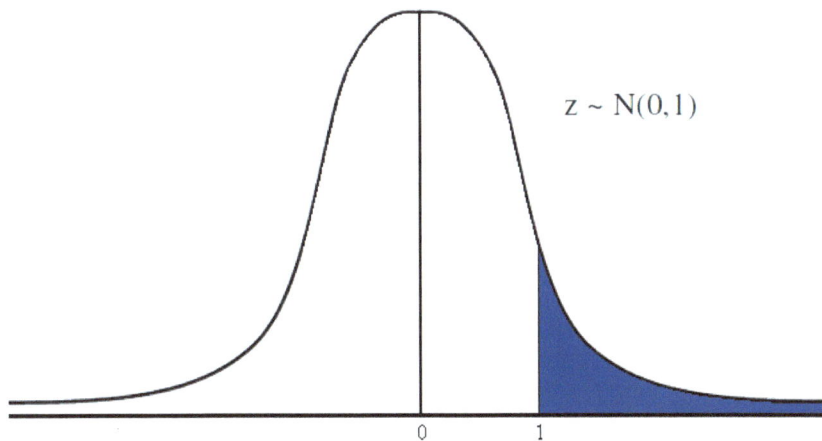

$z \sim N(0,1)$

This version of the z-table gives the area in light blue as 0.3413.
Therefore, $p(0 < z < 1) = 0.3413$.

Hence, $p(z > 1) = 0.5 - p(0 < z < 1)$
$= 0.5 - 0.3413$
$= 0.1587$

Using z-table 2

We look for the z-value '1.00' in the same manner as before. The number at the intersection of the '1.0' row and the '.00' column would give us the area to the right of z = 1.00.

z	.00	.01	.02	.03	.04	.05
0.0	0.5000	0.4960	0.4920	0.4880	0.4840	0.4801
0.1	0.4602	0.4562	0.4522	0.4483	0.4443	0.4404
0.2	0.4207	0.4168	0.4129	0.4090	0.4052	0.4013
0.3	0.3821	0.3783	0.3745	0.3707	0.3669	0.3632
0.4	0.3446	0.3409	0.3372	0.3336	0.3300	0.3264
0.5	0.3085	0.3050	0.3015	0.2981	0.2946	0.2912
0.6	0.2743	0.2709	0.2676	0.2643	0.2611	0.2578
0.7	0.2420	0.2389	0.2358	0.2327	0.2296	0.2266
0.8	0.2119	0.2090	0.2061	0.2033	0.2005	0.1977
0.9	0.1841	0.1814	0.1788	0.1762	0.1736	0.1711
1.0	0.1587	0.1562	0.1539	0.1515	0.1492	0.1469
1.1	0.1357	0.1335	0.1314	0.1292	0.1271	0.1251
1.2	0.1151	0.1131	0.1112	0.1093	0.1075	0.1056
1.3	0.0968	0.0951	0.0934	0.0918	0.0901	0.0885
1.4	0.0808	0.0793	0.0778	0.0764	0.0749	0.0735

This area is 0.1587, which in this instance corresponds to the exact region that we are looking for, so we can therefore say that $p(z > 1) = 0.1587$

Using z-table 3

We again look for the z-value '1.00'. The number in the body of the table that corresponds to a z-value of '1.00' is 0.8413, which represents the entire area to the left of z = 1, made up of the area between z = 1 and z = 0, in addition to the entire left half of the curve.

z	.00	.01	.02	.03	.04
0.0	0.5000	0.5040	0.5080	0.5120	0.5160
0.1	0.5398	0.5438	0.5478	0.5517	0.5557
0.2	0.5793	0.5832	0.5871	0.5910	0.5948
0.3	0.6179	0.6217	0.6255	0.6293	0.6331
0.4	0.6554	0.6591	0.6628	0.6664	0.6700
0.5	0.6915	0.6950	0.6985	0.7019	0.7054
0.6	0.7257	0.7291	0.7324	0.7357	0.7389
0.7	0.7580	0.7611	0.7642	0.7673	0.7704
0.8	0.7881	0.7910	0.7939	0.7967	0.7995
0.9	0.8159	0.8186	0.8212	0.8238	0.8264
1.0	0.8413	0.8438	0.8461	0.8485	0.8508
1.1	0.8643	0.8665	0.8686	0.8708	0.3729
1.2	0.8849	0.8869	0.8888	0.8907	0.3925
1.3	0.9032	0.9049	0.9066	0.9082	0.4099
1.4	0.9192	0.9207	0.9222	0.9236	0.4251

Consider the entire area under the curve. This area is comprised of the light blue region to the left of z = 1 and the dark blue area to the right of z = 1.

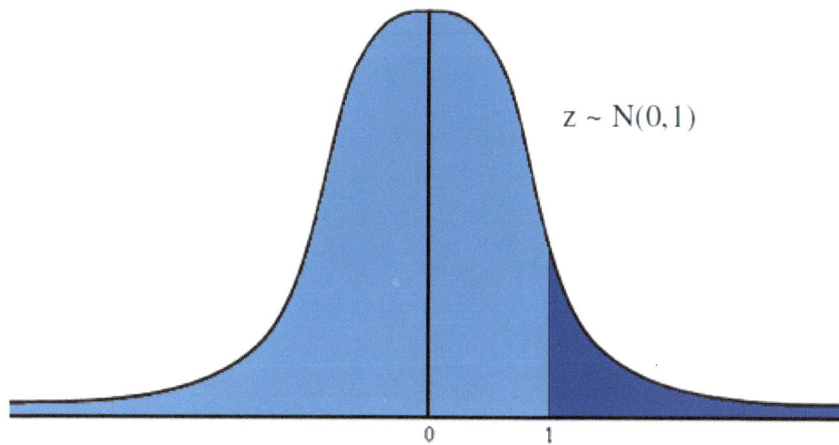

We obtain the dark blue area by taking away the area in light blue from the entire area under the curve, remaining with:

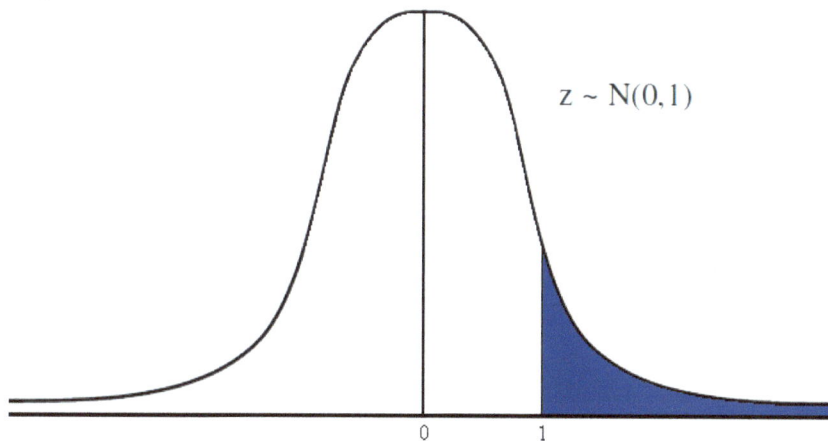

Therefore, $p(z > 1) = 1 - p(z < 1)$
$$= 1 - 0.8413$$
$$= 0.1587$$

So you see, the answer is the same regardless of which of the three tables is used to solve the problem. So, go on now through the following exercises designed to familiarize you with the three versions of the z-table presented.

Exercise 3.1
Using all three versions of the z-table, find:

    (1) $p(z > 1.01)$ (6) $p(z > 3.03)$
    (2) $p(z > 0.51)$ (7) $p(z > 2.00)$
    (3) $p(z > 1.49)$ (8) $p(z > 0.07)$
    (4) $p(z > 0.85)$ (9) $p(z > 1.96)$
    (5) $p(z > 2.72)$ (10) $p(z > 0.66)$

If we wish to find $p(z < -1)$, the area under consideration would be the area to the left of '-1' on the z-curve:

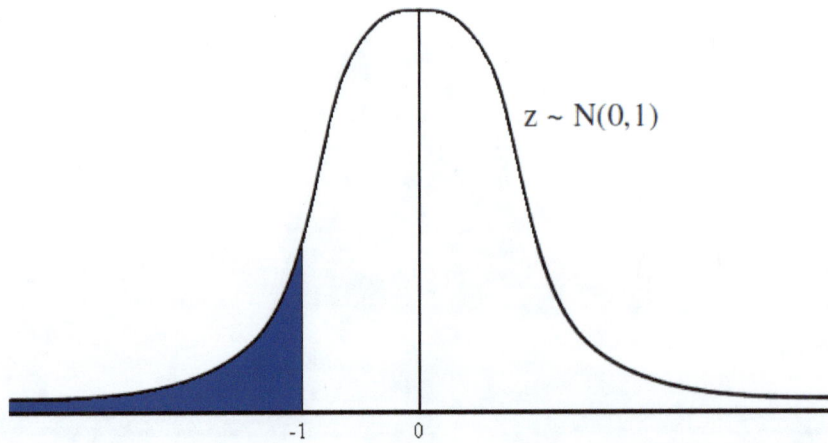

z ~ N(0,1)

Using z-table 1

Consider the entire area to the left of 0 which is equal to 0.5 made up of the area in light blue given by z-table 1 plus the area in dark blue whose value we wish to find.

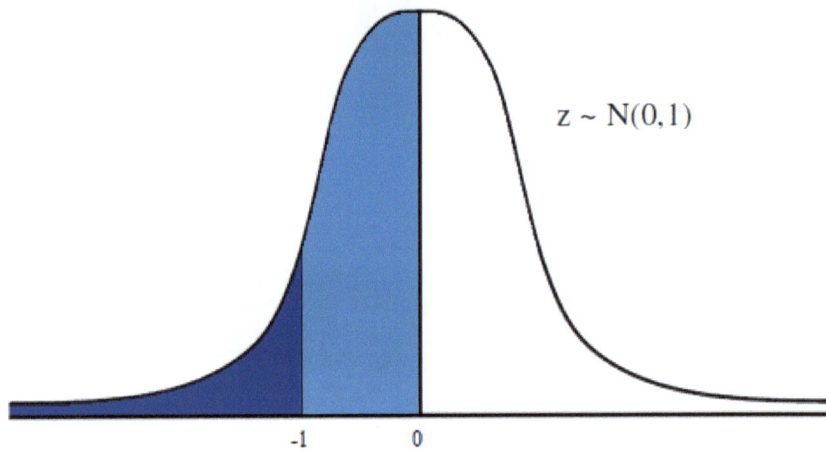

z ~ N(0,1)

When we remove the area in light blue between z = -1 and z = 0, we are left with:

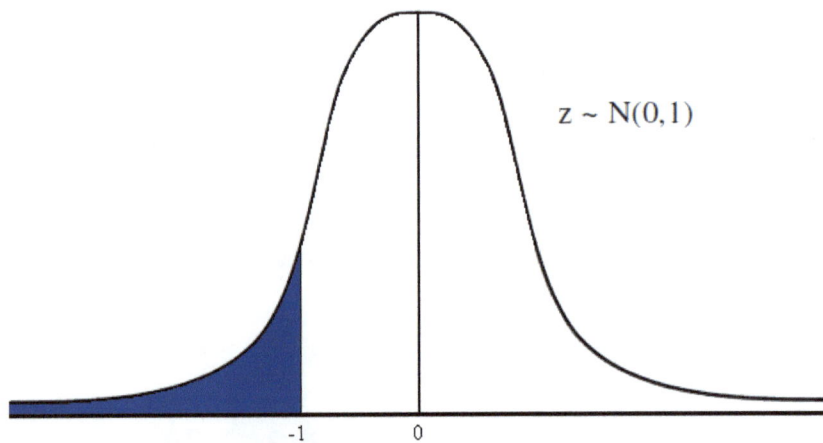

z ~ N(0,1)

In terms of the actual areas: $p(z < -1) = 0.5 - p(-1 < z < 0)$

$$= 0.5 - p(0 < z < 1) \qquad p(-1 < z < 0) = p(0 < z < 1)$$
$$= 0.5 - 0.3413$$
$$= 0.1587$$

Using z-table 2

By virtue of the symmetry property of the normal distribution, the area to the left of z = -1 is very straightforwardly equal to the area to the right of z = 1, which is given directly in this version of the z-table, so we can simply say: $p(z < -1) = p(z > 1)$
$$= 0.1587$$

Using z-table 3

Consider the entire area under the curve comprised of the light blue region to the right of z = -1, as well as the dark blue area to the left of z = -1.

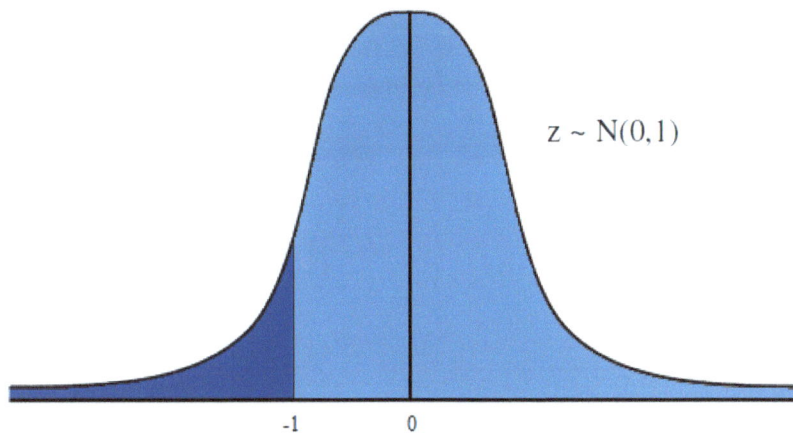

We obtain the required area in dark blue by taking away the light blue area to get:

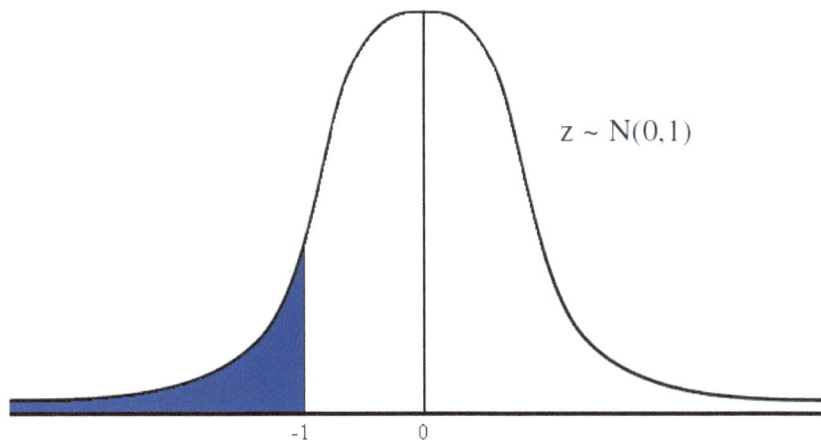

So, using this version of the z-table, we have: $p(z > -1) = 1 - p(z > -1)$
$$= 1 - p(z < 1) \qquad p(z > -1) = p(z < 1)$$
$$= 1 - 0.8413$$
$$= 0.1587$$

Exercise 3.2
Using all three versions of the z-table, find:

    (1) p(z < -1.01)   **(6)** p(z < -3.07)
    (2) p(z < - 0.51)   **(7)** p(z < -2.97)
    (3) p(z < - 1.32)   **(8)** p(z < -1.64)
    (4) p(z < - 0.16)   **(9)** p(z < -0.99)
    (5) p(z < -2.41)   **(10)** p(z < -1.96)

What if we wanted to find p(0.5 < z < 1)? On the z – curve, this area is represented by:

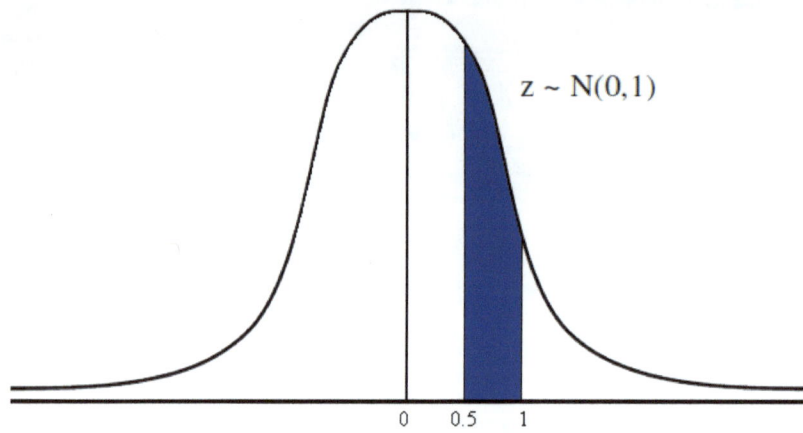

So, what we are really interested in here is the area between 0.5 and 1.0 on the z – curve.

Using z-table 1
Consider the area between 0 and 1 on the standard normal distribution.

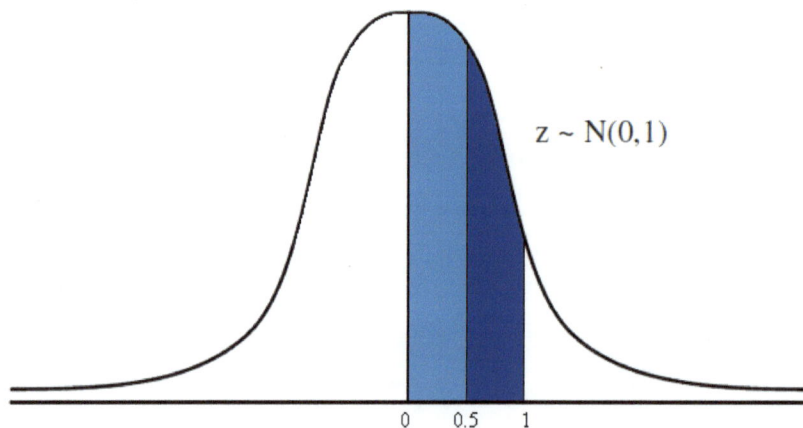

The area in light blue between z = 0 and z = 0.5 is included in the area between z = 0 and z = 1. We therefore get the area between '0.5' and '1' by taking away the area between '0' and '0.5' from the area between '0' and '1', leaving us with:

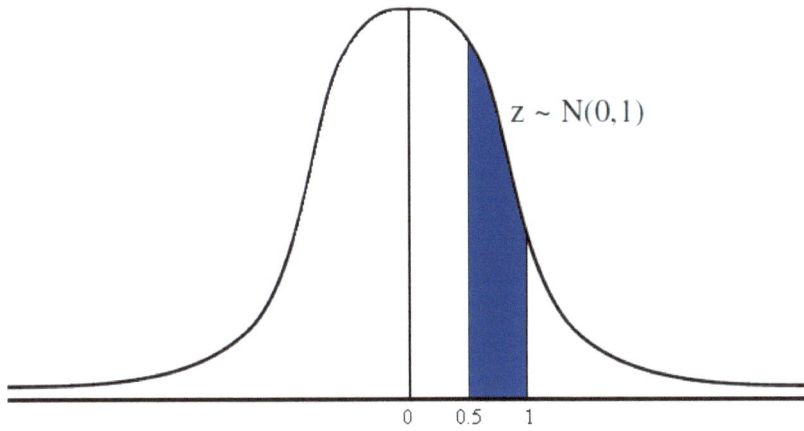

So: $p(0.5 < z < 1) = p(0 < z < 1) - p(0 < z < 0.5)$
$= 0.3413 - 0.1915$
$= 0.1498$

Using z-table 2

Consider the entire area to the right of 0.5:

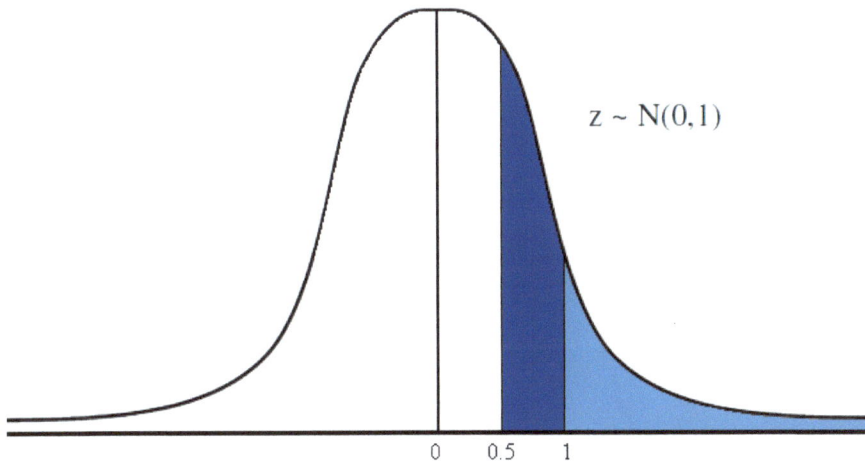

The area to the right of $z = 1$ is included within the area to the right of $z = 0.5$. Therefore, to find the area between 0.5 and 1.0, we take away the area to the right of $z = 1$, to get:

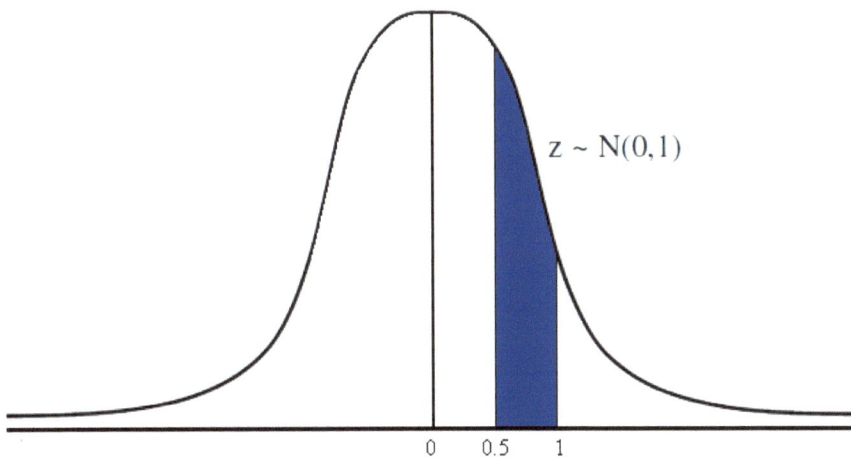

Therefore: $p(0.5 < z < 1) = p(z > 0.5) - p(z > 1)$
$$= 0.3085 - 0.1587$$
$$= 0.1498$$

Using z-table 3

Consider the entire area to the left of z = 1 on the z-curve. The area to the left of z = 0.5 is included in the area to the left of z = 1.

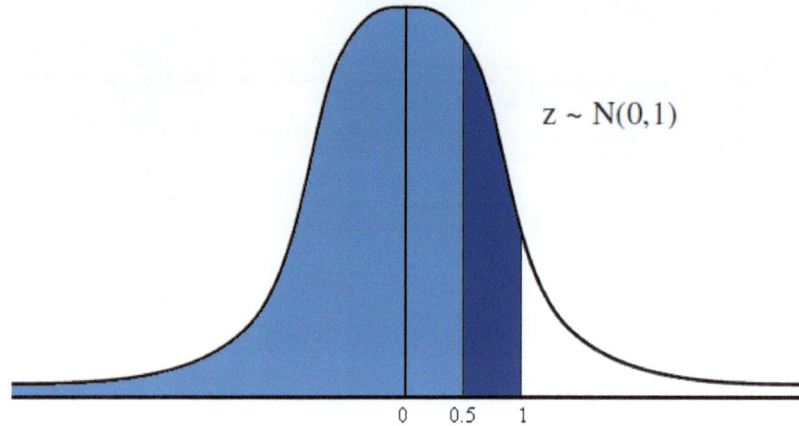

To obtain the dark blue area between '0.5' and '1', we simply remove the light blue area to the left of '0.5', giving us:

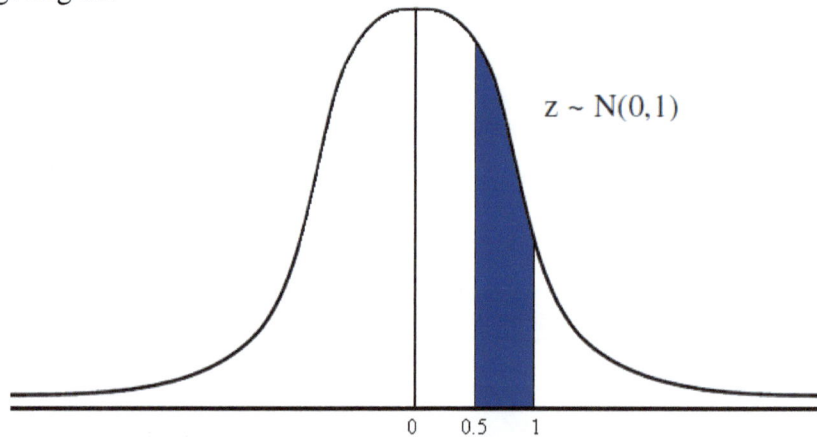

So: $p(0.5 < z < 1) = p(z < 1) - p(z < 0.5)$
$$= 0.8413 - 0.6915$$
$$= 0.1498$$

Exercise 3.3
Using all three versions of the z-table, find:

(1) $p(0.03 < z < 1.03)$ (6) $p(0.36 < z < 2.36)$
(2) $p(0.99 < z < 2.54)$ (7) $p(2.16 < z < 2.84)$
(3) $p(1.33 < z < 2.68)$ (8) $p(0.05 < z < 1.16)$
(4) $p(0.37 < z < 2.06)$ (9) $p(0.72 < z < 1.69)$
(5) $p(1.18 < z < 1.99)$ (10) $p(1.00 < z < 2.00)$

What about p(-1 < z < -0.5)? The required area would be:

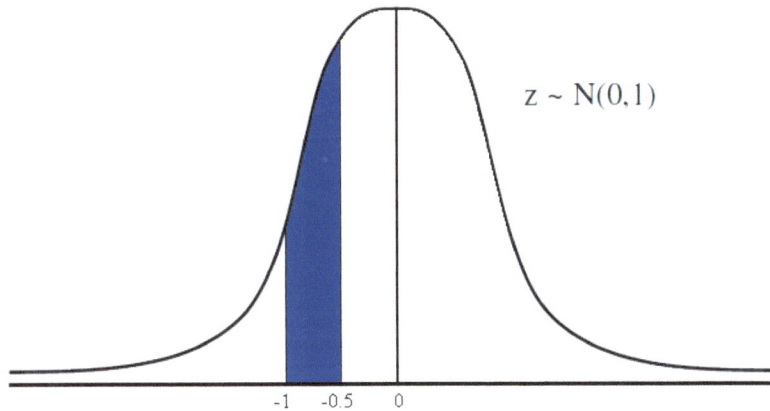

z ~ N(0,1)

-1   -0.5   0

Using z-table 1

Consider first the area between z = -1 and z = 0. This includes the area between z = -0.5 and z = 0.

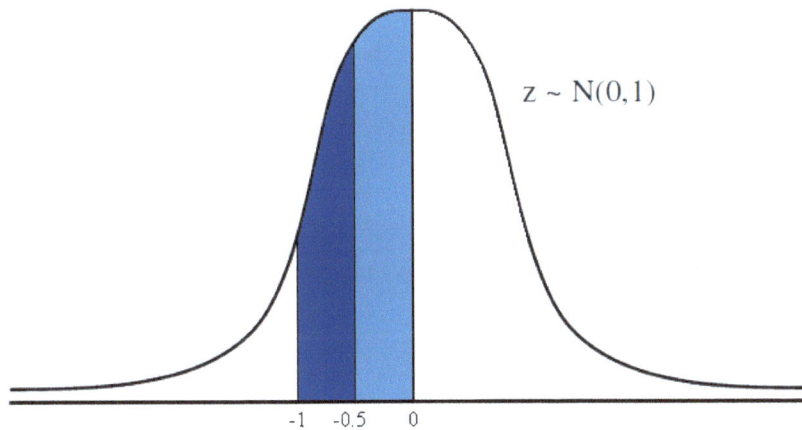

z ~ N(0,1)

-1   -0.5   0

By symmetry this area is exactly equal to the area between z = 0 and z = 1, which is equal to 0.3413. Therefore p(-1 < z < 0) = p(0 < z < 1)
$$= 0.3413$$

Similarly, the area between z = -0.5 and z = 0 is equal to the area between z = 0 and z = 0.5, which is equal to 0.1915. Therefore p(-0.5 < z < 0) = p(0 < z < 0.5)
$$= 0.1915$$

The area between z = -1 and z = -0.5 would now be found by taking away the area between z = -0.5 and z = 0 from the area between z = -1 and z = 0, leaving us with:

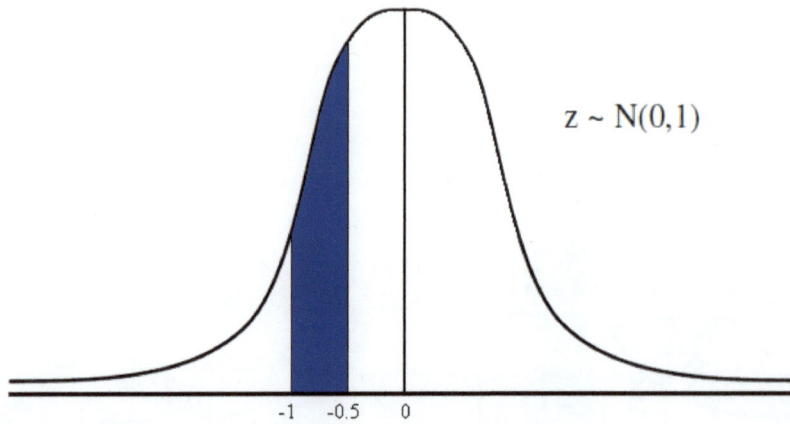

So: $p(-1 < z < -0.5) = p(-1 < z < 0) - p(-0.5 < z < 0)$
$= p(0 < z < 1) - p(0 < z < 0.5)$
$= 0.3413 - 0.1915$
$= 0.1498$

Using z-table 2
Consider first the entire area to the left of $z = -0.5$.

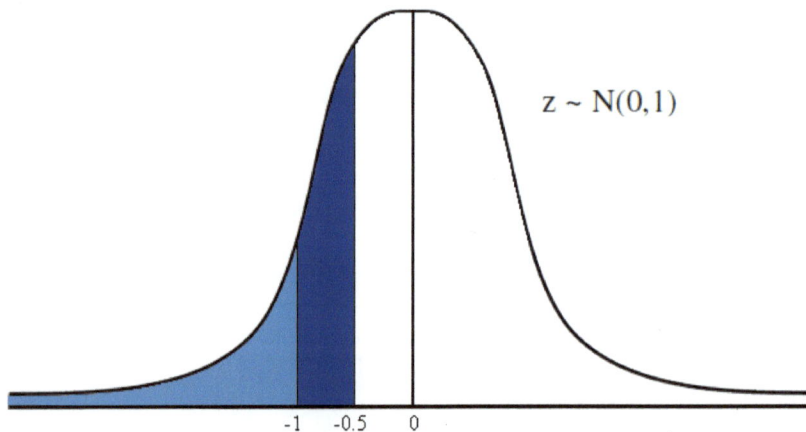

This is the area $p(z < -0.5)$. By symmetry, this area is exactly equal to the area given by $p(z > 0.5)$, which we can quickly find from this version of the z-table as being equal to 0.3085.

The area in light blue to the left of $z = -1$ is denoted by $p(z < -1)$. By symmetry, this area is exactly equal to the area denoted by $p(z > 1)$. This area is equal to 0.1587.

Therefore the area between -1 and -0.5 would be found by taking away the area to the left of $z = -1$, to get:

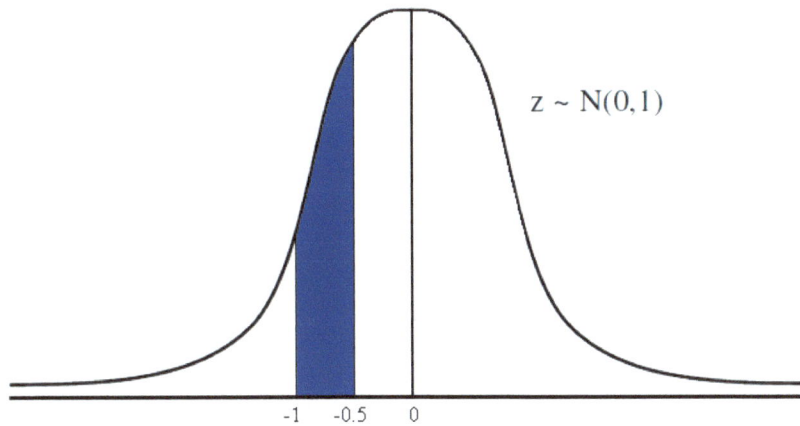

We can therefore say that: p(-1 < z < -0.5) = p(z < -0.5) – p(z < -1)
$$= p(z > 0.5) – p(z > 1)$$
$$= 0.3085 – 0.1587$$
$$= 0.1498$$

Using z-table 3
Consider the area to the right of z = -1. This area is denoted by p(z > -1).

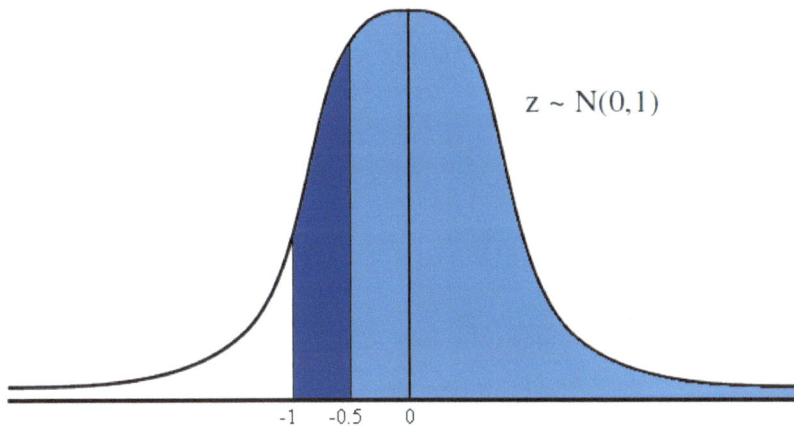

By symmetry, this area is exactly equal to the area to the left of z = 1, denoted by p(z < 1). This area is equal to 0.8413.

The area in light blue is denoted by p(z > - 0.5). By symmetry, this area is exactly equal to the area to the left of z = 0.5, which we can easily find in z-table 3.
Therefore, p(z > - 0.5) = p(z < 0.5)
$$= 0.6915$$

Therefore the area between '-1' and '-0.5' is found by taking away the area to the right of z = -0.5 from the area to the right of z = -1, leaving us with:

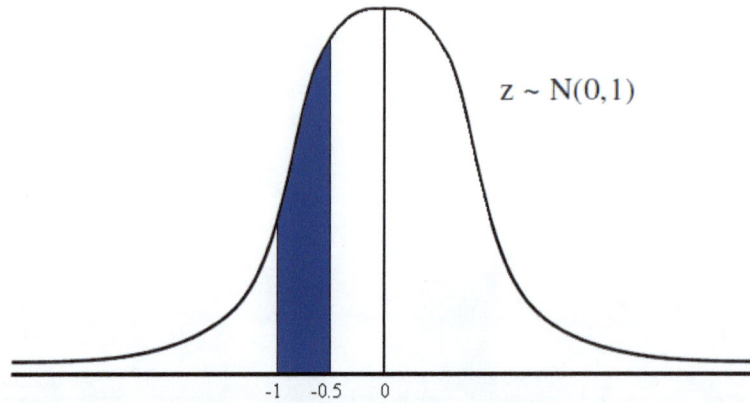

Therefore: $p(-1 < z < -0.5) = p(z > -1) - p(z > -0.5)$
$= p(z < 1) - p(z < 0.5)$
$= 0.8413 - 0.6915$
$= 0.1498$

Exercise 3.4
Using all three versions of the z-table, find:

    (1) $p(-2.65 < z < -0.22)$ (6) $p(-1.03 < z < -0.03)$
    (2) $p(-1.96 < z < -1.00)$ (7) $p(-2.54 < z < -0.99)$
    (3) $p(-2.56 < z < -0.28)$ (8) $p(-1.11 < z < -0.11)$
    (4) $p(-1.87 < z < -0.09)$ (9) $p(-0.97 < z < -0.08)$
    (5) $p(-3.00 < z < -1.53)$ (10) $p(-2.32 < z < -1.22)$

What if we had to find $p(-1 < z < 1)$? The required area would be:

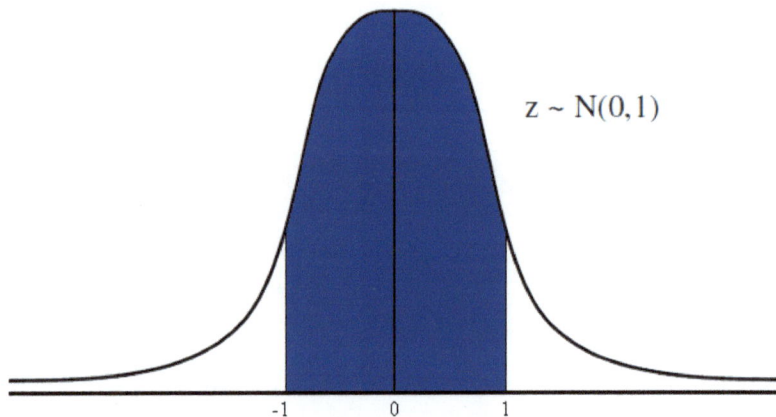

Using z-table 1
The area between $z = 0$ and $z = 1$ is equal to 0.3413. By virtue of the symmetry property of the z-curve, the area between $z = -1$ and $z = 0$ is also exactly equal to 0.3413. So using z-table 1, the area between $z = -1$ and $z = 1$ is found by simply adding the values for the two areas just described: So, $p(-1 < z < 1) = p(-1 < z < 0) + p(0 < z < 1)$
$= 0.3413 + 0.3413$
$= 0.6826$

Using z-table 2

Consider the entire area under the z-curve:

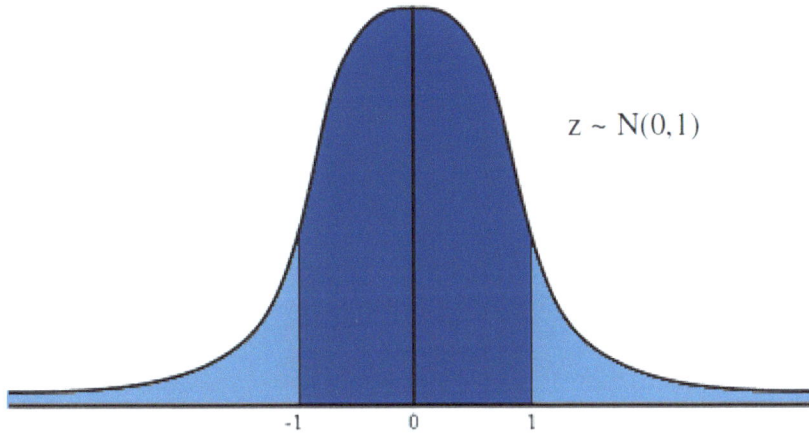

We can obtain the required area in dark blue by taking away the light blue areas in the left and right tails of the curve to get:

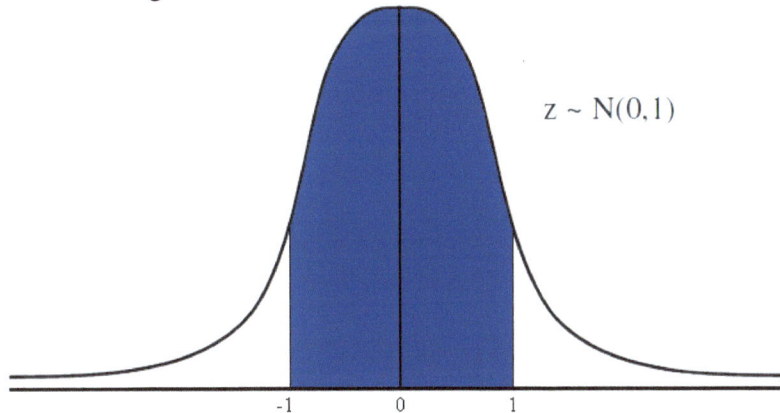

Recall that the total area under the z-curve is equal to 1. We obtain the area to the right of $z = 1$ directly from the z-table. This area is equal to 0.1578. We also know that the area to the left of $z = -1$ is the same as the area to the right of $z = 1$, based on the symmetry property. So this area too, is equal to 0.1578. Therefore we obtain the area between -1 and 1 by simply finding the sum of the area to the left of $z = -1$ and the area to the right of $z = 1$, and subtracting this combined area from the overall area under the curve, which is equal to 1.

So we have:
$$
\begin{aligned}
p(-1 < z < 1) &= 1 - [p(z < -1) + p(z > 1)] \\
&= 1 - [p(z > 1) + p(z > 1)] \\
&= 1 - (0.1587 + 0.1587) \\
&= 1 - 0.3174 \\
&= 0.6826
\end{aligned}
$$

Using z-table 3
Consider the entire area under the z-curve:

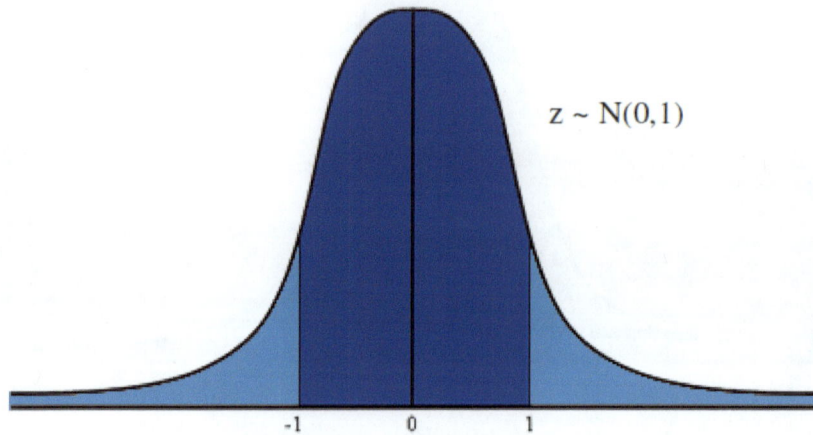

$z \sim N(0,1)$

We need to find the areas in light blue individually, add them, and then subtract the total from 1. Consider first the area in light blue to the left of z = -1. As discussed previously on page 27, this area is given using z-table 3 by: p(z > -1) = 1 - p(z > - 1)

$$= 1 - p(z < 1)$$
$$= 1 - 0.8413$$
$$= 0.1587$$

Consider now the area in light blue to the right of z = 1. As discussed previously on page 25, this area is given using z-table 3 by: p(z > 1) = 1 - p(z < 1)

$$= 1 - 0.8413$$
$$= 0.1587$$

Now, add these two areas: 0.1587 + 0.1587 = 0.3174
Then subtract this result from 1: 1 - 0. 3174 = 0.6826

Therefore, p(-1 < z < 1) = 0.6826

Alternatively (and a little more complicatedly), we can write:
p(-1 < z < 1) = 1 - [{1 - p(z < 1)} + {1 - p(z > -1)}]
$$= 1 - [\{1 - 0.8413\} + \{1 - 0.8413\}]$$
$$= 1 - [\{0.1587\} + \{0.1587\}]$$
$$= 1 - [0.3174]$$
$$= 0.6826$$

Exercise 3.5
Using all three versions of the z-table, find:

(1) p(-0.93 < z < 1.32)	(6) p(-1.63 < z < 1.63)
(2) p(-1.71 < z < 0.93)	(7) p(-0.12 < z < 1.67)
(3) p(-2.30 < z < 0.15)	(8) p(-2.47 < z < 0.54)
(4) p(-0.22 < z < 1.09)	(9) p(-3.04 < z < 2.16)
(5) p(-1.24 < z < 2.11)	(10) p(-0.55 < z < 2.01)

And if we need to find p(z < 1)? The area on the z-curve that concerns us would be:

$$z \sim N(0,1)$$

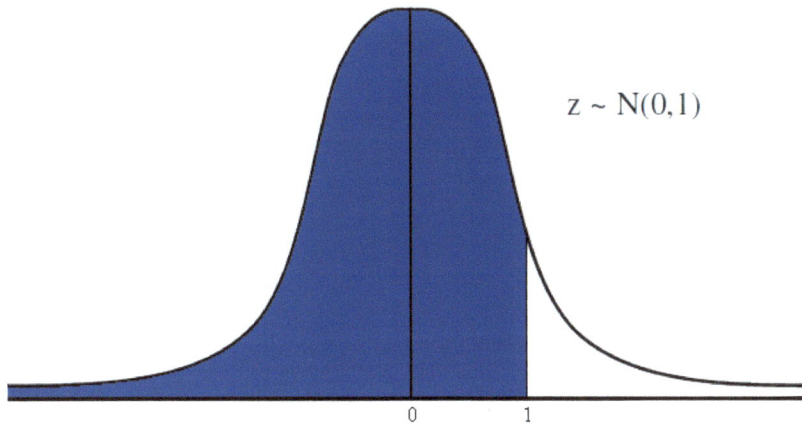

Using z-table 1

The area in the entire left side of the z-curve is equal to 0.5. The area between z = 0 and z = 1 is 0.3413. Therefore the required area in blue is found by simply adding 0.5 to 0.3413.

Therefore: p(z < 1) = 0.5 + p(0 < z < 1)
$$= 0.5 + 0.3413$$
$$= 0.8413$$

Using z-table 2

We can determine the area to the right of z = 1 directly from the z-table. This area is 0.1587. The required area in blue would then be found by taking this area away from 1.

Therefore, p(z < 1) = 1 − p(z > 1)
$$= 1 − 0.1587$$
$$= 0.8413.$$

Using z-table 3

We have it all laid out on a platter here!! The area we are concerned with in this instance is exactly the area that this version of the z-table gives us. So, simply put,
p(z < 1) = 0.8413.

Exercise 3.6
Using all three versions of the z-table, find:

(1) p(z < 2.01)	**(6)** p(z < 3.08)
(2) p(z < 0.45)	**(7)** p(z < 2.69)
(3) p(z < 2.02)	**(8)** p(z < 0.24)
(4) p(z < 0.99)	**(9)** p(z < 1.79)
(5) p(z < 1.45)	**(10)** p(z < 1.08)

And what about if we needed to find p(z > -1)? In this instance, the area on the z-curve that concerns us would be:

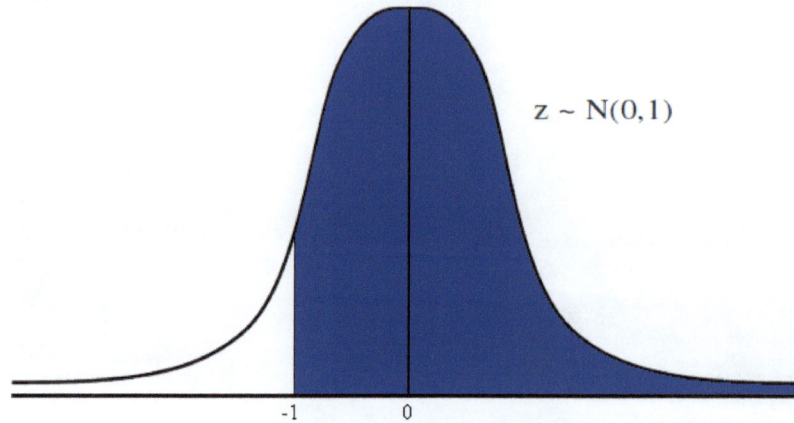

$z \sim N(0,1)$

Using z-table 1

The area in the entire right side of the z-curve is equal to 0.5. By symmetry, we know that the area between z = -1 and z = 0 is equal to the area between z = 0 and z = 1. The table gives this area as 0.3413. Therefore the required area shaded in blue is found by simply adding 0.5 to 0.3413. So: p(z > 1) = 0.5 + p(-1 < z < 0)

$$= 0.5 + p(0 < z < 1)$$
$$= 0.5 + 0.3413$$
$$= 0.8413$$

Using z-table 2

We can get the area to the left of z = -1 by remembering that this is the same as the area to the right of z = 1, which we get directly from the z-table. This area is 0.1587. Therefore the required shaded area in blue is found by subtracting the area to the left of z = -1 from 1.

p(z > -1) = 1 – p(z < -1)
$$= 1 – p(z > 1)$$
$$= 1 – 0.1587$$
$$= 0.8413$$

Using z-table 3

We can get the area to the left of z = 1 directly from this table. This area is equal to 0.8413. By symmetry, the area to the left of z = 1 is exactly equal to the area to the right of z = -1. Therefore: p(z > -1) = p(z < 1)

$$= 0.8413$$

Exercise 3.7

Using all three versions of the z-table, find:

(1) p(z > -1.74)	**(6)** p(z > -0.62)
(2) p(z > -1.96)	**(7)** p(z > -1.33)
(3) p(z > -0.66)	**(8)** p(z > -2.65)
(4) p(z > -2.50)	**(9)** p(z > -1.99)
(5) p(z > - 0.95)	**(10)** p(z > -3.02)

Finding the z-value that corresponds to a particular probability

Very often in Normal Distribution problems, instead of having to find the probability of an event, we may already know that probability. We then need to find the value or values of the random variable associated with the probability of the event in question. In terms of the z-table what this means is that we already know the area under consideration. We then need to find the z-value that would give us that particular area, from which we would go on the find the actual value of the random variable in question. For example, we may have the following situation:

$p(z > x) = 0.1587$. Here, we need to find the z-value that has an area of 0.1587 to its right on the z-curve. The area we seek is:

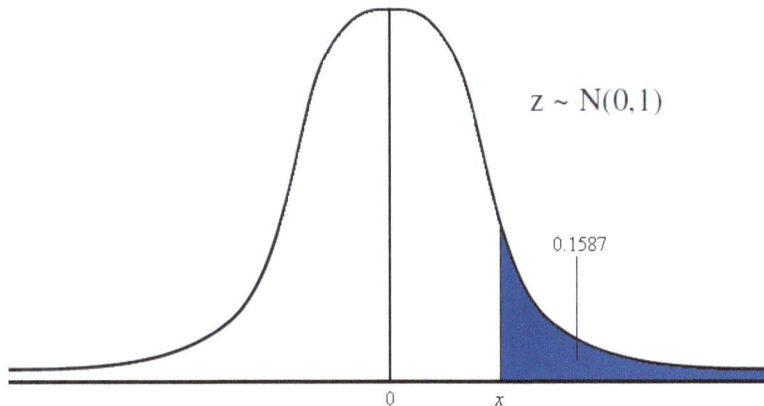

Using z-table 1

This version of the z-table gives the area between $z = 0$ and $z = x$. We only know the area to the right of 'x'. So we find the area between $z = 0$ and $z = x$ by subtracting the area to the right of 'x' from 0.5. We therefore have: $p(0 < z < x) = 0.5 - 0.1587$

$$= 0.3413$$

We now go into the *body* of the z-curve and look for the value 0.3413:

z	.00	.01	.02	.03	.04
0.0	0.0000	0.0040	0.0080	0.0120	0.0160
0.1	0.0398	0.0438	0.0478	0.0517	0.0557
0.2	0.0793	0.0832	0.0871	0.0910	0.0948
0.3	0.1179	0.1217	0.1255	0.1293	0.1331
0.4	0.1554	0.1591	0.1628	0.1664	0.1700
0.5	0.1915	0.1950	0.1985	0.2019	0.2054
0.6	0.2257	0.2291	0.2324	0.2357	0.2389
0.7	0.2580	0.2611	0.2642	0.2673	0.2704
0.8	0.2881	0.2910	0.2939	0.2967	0.2995
0.9	0.3159	0.3186	0.3212	0.3238	0.3264
1.0	0.3413	0.3438	0.3461	0.3485	0.3508
1.1	0.3643	0.3665	0.3686	0.3708	0.3729
1.2	0.3849	0.3869	0.3888	0.3907	0.3925
1.3	0.4032	0.4049	0.4066	0.4082	0.4099
1.4	0.4192	0.4207	0.4222	0.4236	0.4251

The value 0.3413 occurs at the intersection of the '1.0' row and the '.00' column. So the z-value we seek is found by adding '1.0' and '.00'. Therefore $x = 1.00$. Therefore, $p(z > 1.00) = 0.1587$

Using z-table 2

We go into the *body* of the z-curve and find the value 0.1587. We then find the row and the column where the value 0.1587 is the intersection, and the corresponding values for that row and that column would give us the corresponding z-value that we seek.

z	.00	.01	.02	.03	.04	.05
0.0	0.5000	0.4960	0.4920	0.4880	0.4840	0.4801
0.1	0.4602	0.4562	0.4522	0.4483	0.4443	0.4404
0.2	0.4207	0.4168	0.4129	0.4090	0.4052	0.4013
0.3	0.3821	0.3783	0.3745	0.3707	0.3669	0.3632
0.4	0.3446	0.3409	0.3372	0.3336	0.3300	0.3264
0.5	0.3085	0.3050	0.3015	0.2981	0.2946	0.2912
0.6	0.2743	0.2709	0.2676	0.2643	0.2611	0.2578
0.7	0.2420	0.2389	0.2358	0.2327	0.2296	0.2266
0.8	0.2119	0.2090	0.2061	0.2033	0.2005	0.1977
0.9	0.1841	0.1814	0.1788	0.1762	0.1736	0.1711
1.0	0.1587	0.1562	0.1539	0.1515	0.1492	0.1469
1.1	0.1357	0.1335	0.1314	0.1292	0.1271	0.1251
1.2	0.1151	0.1131	0.1112	0.1093	0.1075	0.1056
1.3	0.0968	0.0951	0.0934	0.0918	0.0901	0.0885
1.4	0.0808	0.0793	0.0778	0.0764	0.0749	0.0735

The value 0.1587 is at the intersection of the '1.0' row and the '.00' column. Therefore the z-value we are seeking is '1.00', which we get by simply adding '1.0' to '.00'.
Therefore $x = 1.00$. So, $p(z > 1.00) = 0.1587$

Using z-table 3

The area to the left of $z = x$ is given by $1 - 0.1587 = 0.8413$. We therefore need to go to the body of the z-table and look for the z-value that corresponds to an area of 0.8413.

z	.00	.01	.02	.03	.04
0.0	0.5000	0.5040	0.5080	0.5120	0.5160
0.1	0.5398	0.5438	0.5478	0.5517	0.5557
0.2	0.5793	0.5832	0.5871	0.5910	0.5948
0.3	0.6179	0.6217	0.6255	0.6293	0.6331
0.4	0.6554	0.6591	0.6628	0.6664	0.6700
0.5	0.6915	0.6950	0.6985	0.7019	0.7054
0.6	0.7257	0.7291	0.7324	0.7357	0.7389
0.7	0.7580	0.7611	0.7642	0.7673	0.7704
0.8	0.7881	0.7910	0.7939	0.7967	0.7995
0.9	0.8159	0.8186	0.8212	0.8238	0.8264
1.0	0.8413	0.8438	0.8461	0.8485	0.8508
1.1	0.8643	0.8665	0.8686	0.8708	0.3729
1.2	0.8849	0.8869	0.8888	0.8907	0.3925
1.3	0.9032	0.9049	0.9066	0.9082	0.4099
1.4	0.9192	0.9207	0.9222	0.9236	0.4251

The value 0.8413 occurs at the intersection of the '1.0' row and the '.00' column. Therefore the z-value we are seeking is '1.00', which we get by simply adding '1.0' to '.00'. Therefore $x = 1.00$. So, $p(z > 1.00) = 0.1587$

In attempting the following exercise, where the EXACT probability value cannot be found in the body of the z-table, take the value that is mathematically closest to it.

Exercise 3.8
Using all three versions of the z-table, find the value of x where:

(1) p(z > x) = 0.1515   (6) p(z > x) = 0.0345
(2) p(z > x) = 0.2018   (7) p(z > x) = 0.1000
(3) p(z > x) = 0.0495   (8) p(z > x) = 0.0500
(4) p(z > x) = 0.2810   (9) p(z > x) = 0.3248
(5) p(z > x) = 0.00604   (10) p(z > x) = 0.2735

Whenever the exact probability associated with the value of a random variable cannot be found in the body of the z-table, the value closest to it is used, whether that value is greater than, or less than the probability under consideration

---

Suppose we have a situation where p(z < x) = 0.1587. Graphically, this equates to:

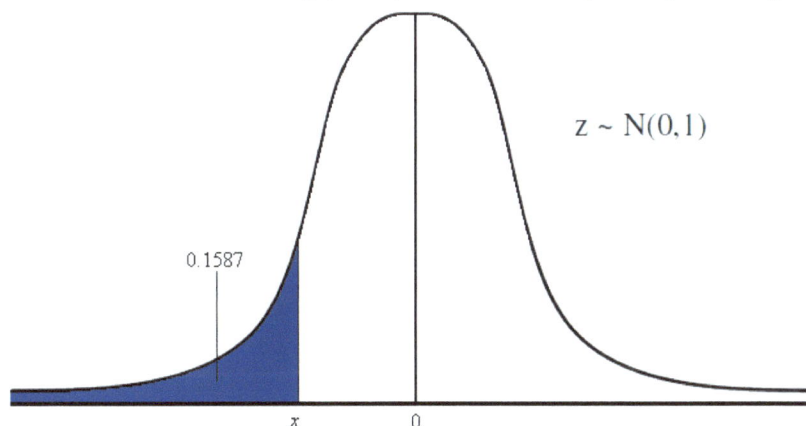

$z \sim N(0,1)$

0.1587

Using z-table 1
The area between z = x and z = 0 is found by subtracting the area to the left of z = x from 0.5.
So we now have: p(x < z < 0) = 0.5 − 0.1587
$$= 0.3413$$
We now go into the body of the z-table looking for 0.3413.

z	.00	.01	.02	.03	.04
0.0	0.0000	0.0040	0.0080	0.0120	0.0160
0.1	0.0398	0.0438	0.0478	0.0517	0.0557
0.2	0.0793	0.0832	0.0871	0.0910	0.0948
0.3	0.1179	0.1217	0.1255	0.1293	0.1331
0.4	0.1554	0.1591	0.1628	0.1664	0.1700
0.5	0.1915	0.1950	0.1985	0.2019	0.2054
0.6	0.2257	0.2291	0.2324	0.2357	0.2389
0.7	0.2580	0.2611	0.2642	0.2673	0.2704
0.8	0.2881	0.2910	0.2939	0.2967	0.2995
0.9	0.3159	0.3186	0.3212	0.3238	0.3264
1.0	0.3413	0.3438	0.3461	0.3485	0.3508
1.1	0.3643	0.3665	0.3686	0.3708	0.3729
1.2	0.3849	0.3869	0.3888	0.3907	0.3925
1.3	0.4032	0.4049	0.4066	0.4082	0.4099
1.4	0.4192	0.4207	0.4222	0.4236	0.4251

The value 0.3413 occurs at the intersection of the '1.0' row and the '.00' column. So the z-value that we seek is found by adding '1.0' and '.00'. '1.00' + '.00' = '1.00'. Remember however that the z-table are using is right-sided, while our 'x' value is on the left side of the z-curve. Therefore the only possible value for x would be -1.00. Therefore $x = -1$

Using z-table 2

We have to find the value x on the z-table that has an area of 0.1587 to its left. Since this value lies on the left side of the z-table, and the z-table gives values from the right side, we again invoke the symmetry characteristic of the z-curve, going into the body of the z-table to find value 0.1587 and its corresponding z-value.

z	.00	.01	.02	.03	.04	.05
0.0	0.5000	0.4960	0.4920	0.4880	0.4840	0.4801
0.1	0.4602	0.4562	0.4522	0.4483	0.4443	0.4404
0.2	0.4207	0.4168	0.4129	0.4090	0.4052	0.4013
0.3	0.3821	0.3783	0.3745	0.3707	0.3669	0.3632
0.4	0.3446	0.3409	0.3372	0.3336	0.3300	0.3264
0.5	0.3085	0.3050	0.3015	0.2981	0.2946	0.2912
0.6	0.2743	0.2709	0.2676	0.2643	0.2611	0.2578
0.7	0.2420	0.2389	0.2358	0.2327	0.2296	0.2266
0.8	0.2119	0.2090	0.2061	0.2033	0.2005	0.1977
0.9	0.1841	0.1814	0.1788	0.1762	0.1736	0.1711
1.0	0.1587	0.1562	0.1539	0.1515	0.1492	0.1469
1.1	0.1357	0.1335	0.1314	0.1292	0.1271	0.1251
1.2	0.1151	0.1131	0.1112	0.1093	0.1075	0.1056
1.3	0.0968	0.0951	0.0934	0.0918	0.0901	0.0885
1.4	0.0808	0.0793	0.0778	0.0764	0.0749	0.0735

The value from the right-sided table is 1.00, as in the last example, but since our x is on the left side of the table, then our z-value in this instance is -1. Therefore, $x = -1$.

Using z-table 3

The area to the right of $z = x$ is equal to $1 - 0.1587 = 0.8413$. We now go into the body of the z-table to find the z-value that corresponds to a value of 0.8413.

z	.00	.01	.02	.03	.04
0.0	0.5000	0.5040	0.5080	0.5120	0.5160
0.1	0.5398	0.5438	0.5478	0.5517	0.5557
0.2	0.5793	0.5832	0.5871	0.5910	0.5948
0.3	0.6179	0.6217	0.6255	0.6293	0.6331
0.4	0.6554	0.6591	0.6628	0.6664	0.6700
0.5	0.6915	0.6950	0.6985	0.7019	0.7054
0.6	0.7257	0.7291	0.7324	0.7357	0.7389
0.7	0.7580	0.7611	0.7642	0.7673	0.7704
0.8	0.7881	0.7910	0.7939	0.7967	0.7995
0.9	0.8159	0.8186	0.8212	0.8238	0.8264
1.0	0.8413	0.8438	0.8461	0.8485	0.8508
1.1	0.8643	0.8665	0.8686	0.8708	0.3729
1.2	0.8849	0.8869	0.8888	0.8907	0.3925
1.3	0.9032	0.9049	0.9066	0.9082	0.4099
1.4	0.9192	0.9207	0.9222	0.9236	0.4251

The value 0.8413 occurs at the intersection of the '1.0' row and the '.00' column. So the z-value that we seek is found by adding '1.0' and '.00' to get '1.00'. But recall that our 'x' lies on the left side of the z-curve, so that our z-value is therefore negative, equal to -1.00. Therefore, $x = -1$.

Exercise 3.9

Using all three versions of the z-table, find the value of x where:

     (**1**) $p(z < x) = 0.1515$ (**6**) $p(z < x) = 0.1678$

     (**2**) $p(z < x) = 0.2153$ (**7**) $p(z < x) = 0.3944$

     (**3**) $p(z < x) = 0.0505$ (**8**) $p(z < x) = 0.0012$

     (**4**) $p(z < x) = 0.00734$ (**9**) $p(z < x) = 0.4124$

     (**5**) $p(z < x) = 0.4920$ (**10**) $p(z < x) = 0.2643$

What if $p(z > x) = 0.6587$? What is the difference between this situation and the one where $p(z > x) = 0.1587$? The difference is that in the first case the value of the area under consideration is greater than 0.5. And what exactly is the significance of '0.5'? Recall that the total area under the z-curve is equal to 1. Recall also that the z-curve is symmetric, which translates graphically to each side of the graph having the same area of 0.5. So we have an area of 0.5 on the left side, and an area of 0.5 on the right side of the curve as well. This means that if $p(z > x)$ has a value greater than 0.5, then x would be on the *left* side of the z-curve as opposed to being on the right side in the case where $p(z > x)$ is less than 0.5, as in the case where $p(z > x) = 0.1587$. Therefore, considering again the case where $p(z > x) = 0.6587$, we have the following graphical situation:

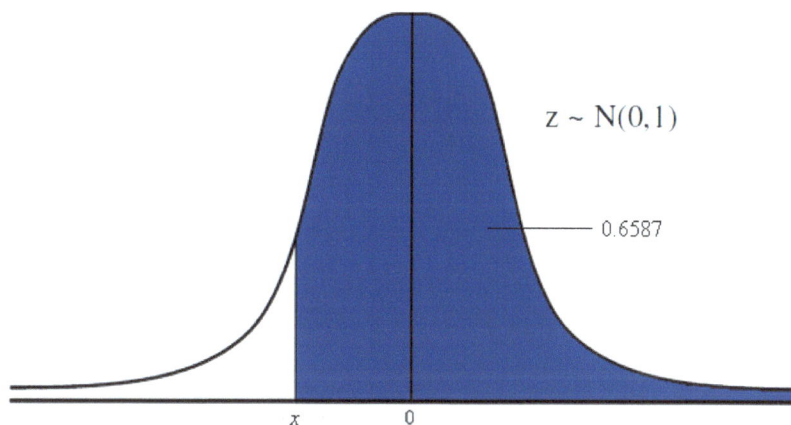

$z \sim N(0,1)$

0.6587

x 0

Using z-table 1

The area between $z = x$ and $z = 0$ is found by subtracting 0.5 from 0.6587. Therefore, $p(x < z < 0) = 0.1587$. Recall that the z-tables we are using are all right-sided. Since our 'x' lies on the left side of the z-curve, we need to invoke the symmetry property. We find the positive z-value that corresponds to an area of 0.1587 between itself and $z = 0$. We go into the body of the z-table and look for the value 0.1587, or the value that is closest to it.

z	.00	.01	.02
0.0	0.0000	0.0040	0.0080
0.1	0.0398	0.0438	0.0478
0.2	0.0793	0.0832	0.0871
0.3	0.1179	0.1217	0.1255
0.4	0.1554	0.1591	0.1628
0.5	0.1915	0.1950	0.1985
0.6	0.2257	0.2291	0.2324
0.7	0.2580	0.2611	0.2642
0.8	0.2881	0.2910	0.2939
0.9	0.3159	0.3186	0.3212

The value in the body of the z-table that is closest to 0.1587 is 0.1591. The right-sided z-value that corresponds to this area is 0.41, but since our 'x' is on the left side of the z-curve, we must use the corresponding left-sided value for 0.41, which would of course be -0.41.

Therefore $x = -0.41$. So, we have, finally, that $p(z > -0.41) = 0.6587$.

Using z-table 2

The shaded area in blue automatically implies an area of $1 - 0.6587 = 0.3413$ to the left of x. We therefore need to find the z-value which has an area of 0.3413 to its left. Recall that the z-table only gives areas for z-values on the right side of the curve. We have here a value x on the left side of the curve, so we look for the positive value of z that has an area of 0.3413 to its right. We look within the body of the z-curve for the value 0.3413. Because the exact value 0.3413 does not occur in the body of the z-table, we take the value closest to it. This value is 0.3409.

z	.00	.01	.02
0.0	0.5000	0.4960	0.4920
0.1	0.4602	0.4562	0.4522
0.2	0.4207	0.4168	0.4129
0.3	0.3821	0.3783	0.3745
0.4	0.3446	0.3409	0.3372
0.5	0.3085	0.3050	0.3015
0.6	0.2743	0.2709	0.2676
0.7	0.2420	0.2389	0.2358
0.8	0.2119	0.2090	0.2061
0.9	0.1841	0.1814	0.1788

We see from the table that the value 0.3409 lies at the intersection of the '0.4' row and the '.01' column. The z-value is found by simply adding these two values. 0.4 + .01 gives us a z-value of 0.41. This z-value of 0.41 gives us the area of 0.3413 that we had before, but remember that x lies on the *left* side of the z-curve. Therefore $x = -0.41$.

Using z-table 3

This z-table will immediately give us a right-sided z-value which has a total area to the left of 0.6587. In the body of the table, the closet value to 0.6587 is 0.6591. The right-sided z-value that corresponds to this value of 0.6591 is 0.41.

z	.00	.01	.02
0.0	0.5000	0.5040	0.5080
0.1	0.5398	0.5438	0.5478
0.2	0.5793	0.5832	0.5871
0.3	0.6179	0.6217	0.6255
0.4	0.6554	0.6591	0.6628
0.5	0.6915	0.6950	0.6985
0.6	0.7257	0.7291	0.7324
0.7	0.7580	0.7611	0.7642
0.8	0.7881	0.7910	0.7939
0.9	0.8159	0.8186	0.8212

But since our 'x' is on the left side of the z-curve, we must use the corresponding left-sided value for 0.41. This value is -0.41. Therefore, $x = -0.41$.

Exercise 3.10

Using all three versions of the z-table, find x where:

(1) $p(z > x) = 0.7654$
(2) $p(z > x) = 0.6$
(3) $p(z > x) = 0.85$
(4) $p(z > x) = 0.55$
(5) $p(z > x) = 0.9125$

(6) $p(z > x) = 0.6385$
(7) $p(z > x) = 0.7245$
(8) $p(z > x) = 0.8264$
(9) $p(z > x) = 0.7148$
(10) $p(z > x) = 0.9239$

What if we need to find $p(z < x) = 0.6587$? As in the example above, we have a situation where the area under consideration is greater than 0.5. The area therefore would be comprised of some portion of both halves of the standard normal curve. We would have the entire left side of the curve covered, in addition to a little extra on the right side. The 'extra' amount would be $0.6587 - 0.5 = 0.1587$.

Firstly, the area we are considering would be:

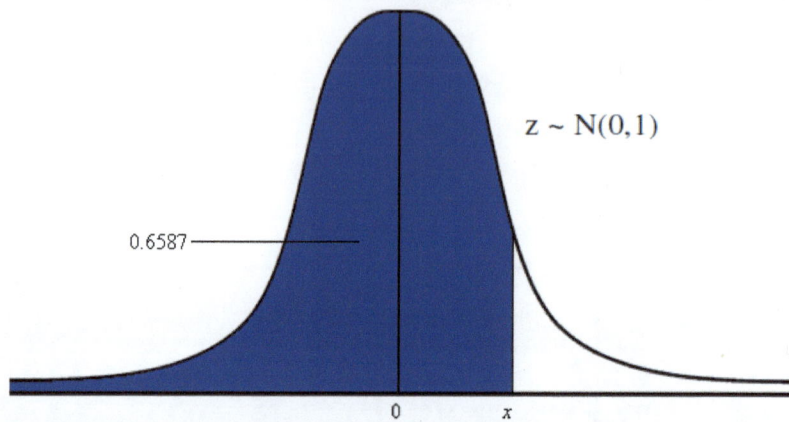

z ~ N(0,1)

0.6587

0   x

Using z-table 1

The area between z = 0 and z = x is found by subtracting 0.5 from 0.6587. Therefore, $p(0 < z < x) = 0.1587$. We now go to the table to find the z-value that corresponds to an area of 0.1587. We go into the body of the z-table and look for the value 0.1587, or the value closest to it.

z	.00	.01	.02
0.0	0.0000	0.0040	0.0080
0.1	0.0398	0.0438	0.0478
0.2	0.0793	0.0832	0.0871
0.3	0.1179	0.1217	0.1255
0.4	0.1554	0.1591	0.1628
0.5	0.1915	0.1950	0.1985
0.6	0.2257	0.2291	0.2324
0.7	0.2580	0.2611	0.2642
0.8	0.2881	0.2910	0.2939
0.9	0.3159	0.3186	0.3212

The value in the body of the z-table that is closest to 0.1587 is 0.1591. The z-value that corresponds to this area is 0.41, therefore $x = 0.41$.

Using z-table 2

We have an area to the right of x of $1 - 0.6587 = 0.3413$. Therefore we need to find x, the z-value with an area of 0.3413 to its right. We go to the body of the z-table, looking to find the value that is closest to 0.3413, if not 0.3413 itself.

z	.00	.01	.02
0.0	0.5000	0.4960	0.4920
0.1	0.4602	0.4562	0.4522
0.2	0.4207	0.4168	0.4129
0.3	0.3821	0.3783	0.3745
0.4	0.3446	0.3409	0.3372
0.5	0.3085	0.3050	0.3015
0.6	0.2743	0.2709	0.2676
0.7	0.2420	0.2389	0.2358
0.8	0.2119	0.2090	0.2061
0.9	0.1841	0.1814	0.1788

The value in the body of the z-table that is closest to 0.3413 is 0.3409. So to get the first decimal place in our z-value x, we go across to the left, to find '0.4'. We get the second decimal place by going up to find '.01'. So the resulting z-value is determined by adding '0.4' and '.01' to get '0.41'. So therefore our z-value is 0.41. Therefore $x = 0.41$

Using z-table 3

This is a situation where we can get the z-value we are seeking in ONE step. The table will immediately give us the z-value that has a total area of 0.6587 to its left. All we need to do is look in the table for the value 0.6587 or the value closest to it.

z	.00	.01	.02
0.0	0.5000	0.5040	0.5080
0.1	0.5398	0.5438	0.5478
0.2	0.5793	0.5832	0.5871
0.3	0.6179	0.6217	0.6255
0.4	0.6554	0.6591	0.6628
0.5	0.6915	0.6950	0.6985
0.6	0.7257	0.7291	0.7324
0.7	0.7580	0.7611	0.7642
0.8	0.7881	0.7910	0.7939
0.9	0.8159	0.8186	0.8212

This value is 0.6591, with a corresponding z-value of 0.41. Therefore, $x = 0.41$.

Exercise 3.11

Using all three versions of the z-table, find x where:

    (1) $p(z < x) = 0.7654$ **(6)** $p(z < x) = 0.6358$
    (2) $p(z < x) = 0.6$ **(7)** $p(z < x) = 0.9367$
    (3) $p(z < x) = 0.5341$ **(8)** $p(z < x) = 0.7447$
    (4) $p(z < x) = 0.8435$ **(9)** $p(z < x) = 0.5943$
    (5) $p(z < x) = 0.9$ **(10)** $p(z < x) = 0.8211$

The p-value

One of the most straightforward, and yet, most misunderstood concepts pertaining to the Normal Distribution is that of the p-value. We have already encountered the concept of the standardized z-value. Allied with the standardized z-value is the concept of the p-value. Every z-value has a corresponding p-value. Similarly, every p-value has an associated z-value.

The p-value of a standardized z-value is the proportion of data that is more extreme than the z-score under consideration.

Consider a standardized z-score of 1.53. On the standard normal curve, we have:

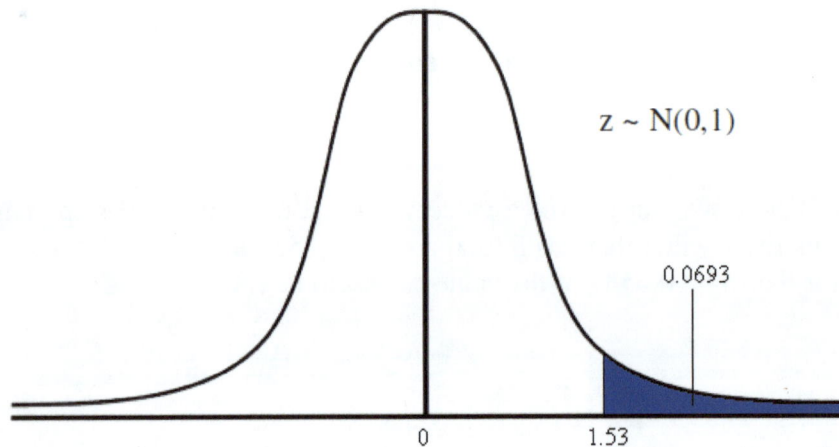

The area of 0.0693 to the extreme right of 1.53 is its p-value. This value represents the probability that a randomly chosen z-score would be more extreme than 1.53. We may also express this p-value as the proportion of data that is more extreme than 1.53.

Consider also a standardized z-score of -1.53. On the standard normal curve, we have:

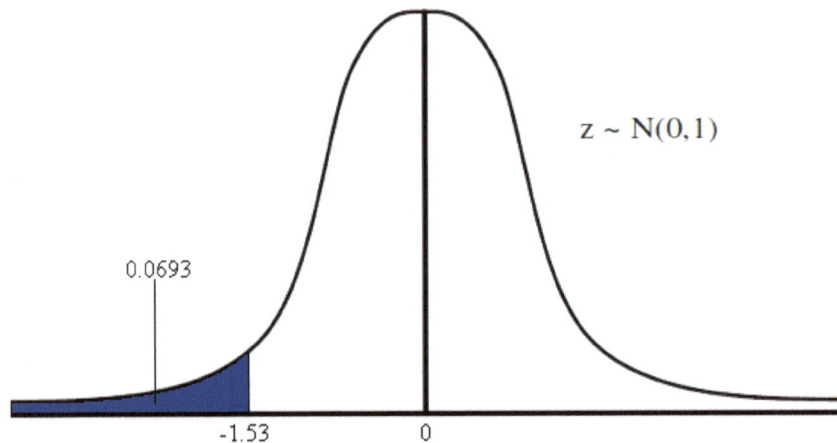

The area of 0.0693 to the extreme left of -1.53 is its p-value. This value represents the probability that a randomly chosen z-score would be more extreme than -1.53. We may also express this p-value as the proportion of data that is more extreme than -1.53.

Note that the standardized z-scores of -1.53 and 1.53 both have an identical p-value of 0.0693. This is because of the symmetry property of the Standard Normal Distribution. When we say that a randomly selected z-value is more extreme than a given standardized z-score, we are in fact saying that it is further away from the mean than the standardized z-score under consideration. In the case of a positive z-score (for example 1.53), the randomly selected z-score is greater than the standardized z-score under consideration. In the case of a negative z-score (for example -1.53), the randomly selected z-score would be less than the standardized z-score under consideration.

The p-value represents a probability, or a proportion. On the Standard Normal Distribution curve it is expressed as an area, and as such the p-value is ALWAYS positive (even when its associated z-value is negative).

The p-value is ALWAYS positive!

Exercise 3.12
Using all three versions of the standardized z-table, determine the p-values of the following standardized z-values:

(1) 0.64	**(6)** -1.45
(2) 1.35	**(7)** -1.28
(3) -2.01	**(8)** 2.57
(4) 1.06	**(9)** 1.81
(5) -0.97	**(10)** -0.64

Determining the standardized z-value using the p-value

Suppose we already know the p-value of a standardized z-score, but we do not know the z-score itself. We know only the sign of the z-score (whether it is 'positive' or 'negative'). In this situation we can use the z-table to obtain the z-score from the p-value.

Consider a situation where we know that the p-value is 0.0693, and that the corresponding standardized z-value is positive. Represented graphically, what we have is:

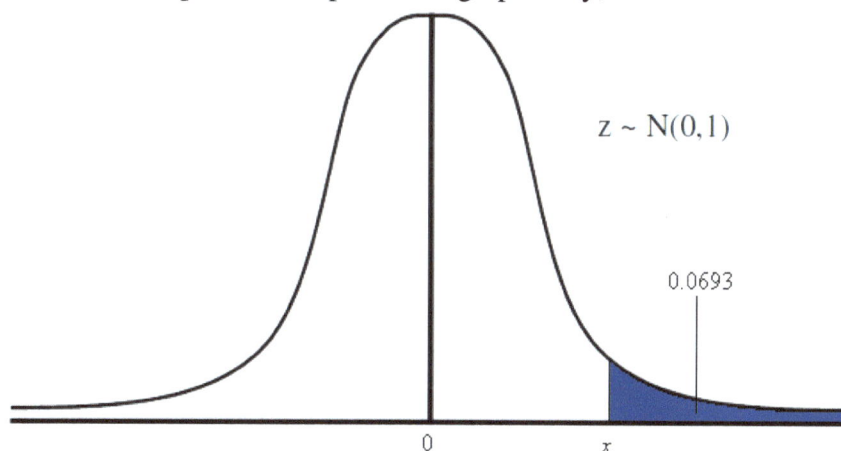

We need to go into the *body* of the z-table to find the positive z-value '*x*' that corresponds to a p-value of 0.0693. This value is 1.53. So, $x = 1.53$.

Consider now the case where we know that the p-value is 0.0693 and that the corresponding z-value is negative. Graphically, we now have:

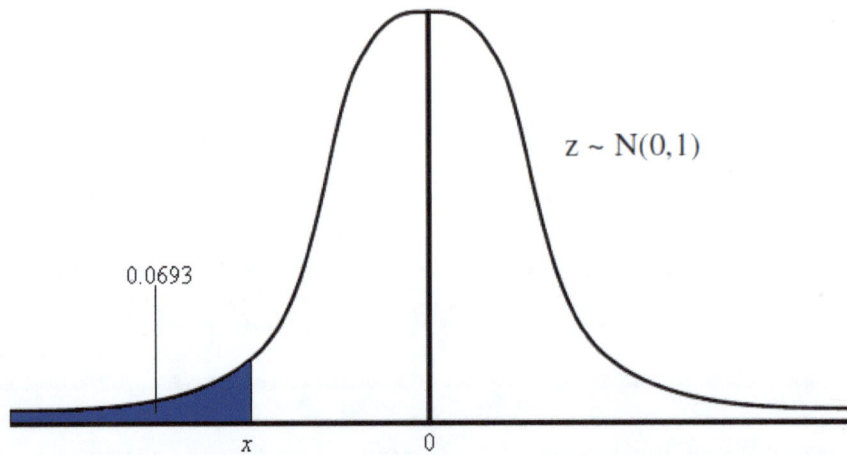

As before, we need to go into the *body* of the z-table to find the negative z-value '*x*' that corresponds to a p-value of 0.0693. Of course, the table will give us the positive value of *z*, which is 1.53. So, we invoke the symmetry property of the z-curve, and conclude that *x* = -1.53.

Exercise 3.13

Using all three versions of the standardized z-table, determine the standardized z-values that correspond to the following p-values:

(1) 0.05 *(-ve z-value)*

(2) 0.093 *(+ve z-value)*

(3) 0.145 *(+ve z-value)*

(4) 0.400 *(-ve z-value)*

(5) 0.025 *(-ve z-value)*

(6) 0.01 *(+ve z-value)*

(7) 0.005 *(-ve z-value)*

(8) 0.035 *(-ve z-value)*

(9) 0.2364 *(+ve z-value)*

(10) 0.13 *(+ve z-value)*

Never attempt a Normal Distribution problem without the aid of the appropriate diagrams!! It is a silly and contemptuous error, which can prove quite costly in an exam situation. The diagram makes it unambiguously clear exactly which areas and which values are under consideration. No matter how confident you are in your ability to solve Normal Distribution problems, *always use diagrams!!*

Example 3.1

The amount of milk produced by a cow on a ranch in Texas each week is normally distributed with a mean of 362.4 pounds and a standard deviation of 52.6 pounds.

(a) If a cow is randomly chosen from this farm and milked for a week, what is the probability that it would produce more than 400 pounds of milk?

(b) The managers of the ranch decide to send the lesser producing cows to be used for production of meat. They determine that cows in the lowest 15 percent of the milk-producers will be used for meat production. What is the level of milk production below which a cow would be sent to be used for meat production?

Solution

(a) The first step in this, as in ALL problems involving statistical distributions, is to define the random variable. So:

Let X be the random variable 'amount of milk in pounds produced by a randomly chosen cow on a ranch in Texas.'

We know the variable is normally distributed, so we can now write:

$X \sim N(362.4, 52.6^2)$

So, we are now required to find: $P(X > 400)$.

The next step, which you should employ in ALL problems involving the normal distribution, is to sketch a graph of the situation:

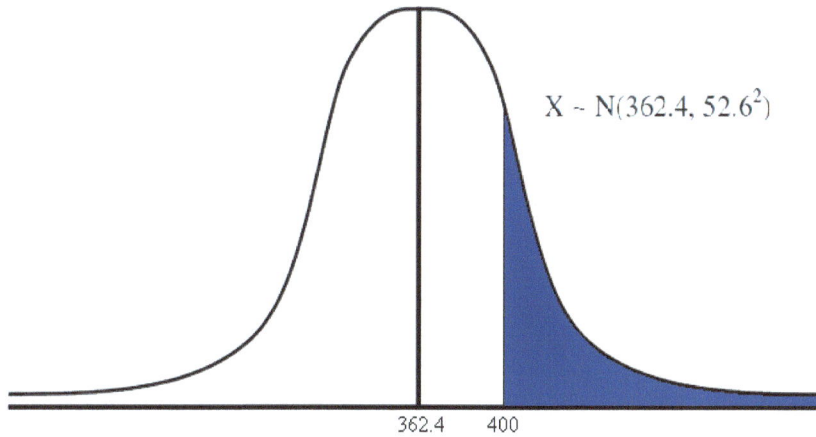

$X \sim N(362.4, 52.6^2)$

362.4   400

We now need to standardize our values, to express our '400' in terms of the number of standard deviations that it lies away from the mean of '362.4'

Standardizing: $z_{400} = \dfrac{400 - 362.4}{56.2}$

$= 0.67$

Therefore, on the standard normal distribution curve, the value lies 0.67 standard deviations away from the mean of 362.4, where the standard deviation is 56.2.

So, in terms of the standard normal curve, we have:

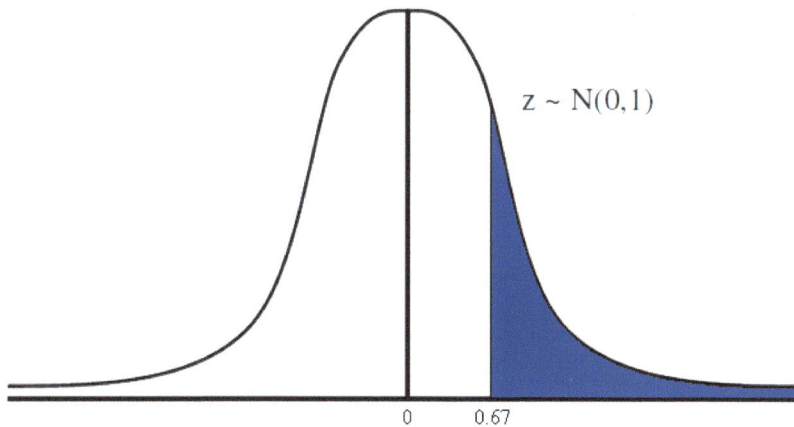

$z \sim N(0,1)$

0   0.67

So, $P(X > 400) = P(z > \dfrac{400 - 362.4}{56.2})$

$\qquad\qquad\qquad = P(z > 0.67)$

$\qquad\qquad\qquad = 0.2514$

Therefore, the probability that a randomly chosen cow from this farm produces more than 400 pounds of milk is 0.2514.

(b) We have already defined the random variable in part (a), so we can start straightaway with the graphical depiction of the situation at hand:

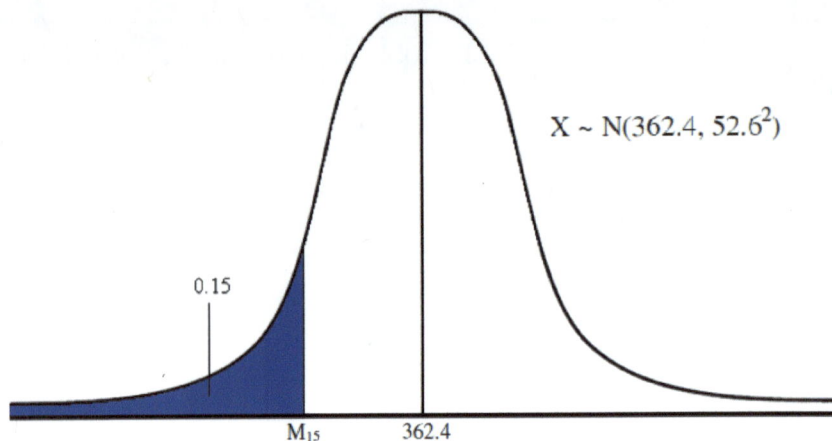

Let M_{15} be the level of milk production below which the lowest fifteen percent of milk producing cows produce. Remember that the normal distribution assumes that the data is arranged in ascending order (from the smallest value to the greatest value), so the shaded region in blue to the left of M_{15} represents the lowest milk producers, up to the first fifteen percent. So, $P(X < M_{15}) = 0.15$

We now need to express this in terms of the standard normal distribution:

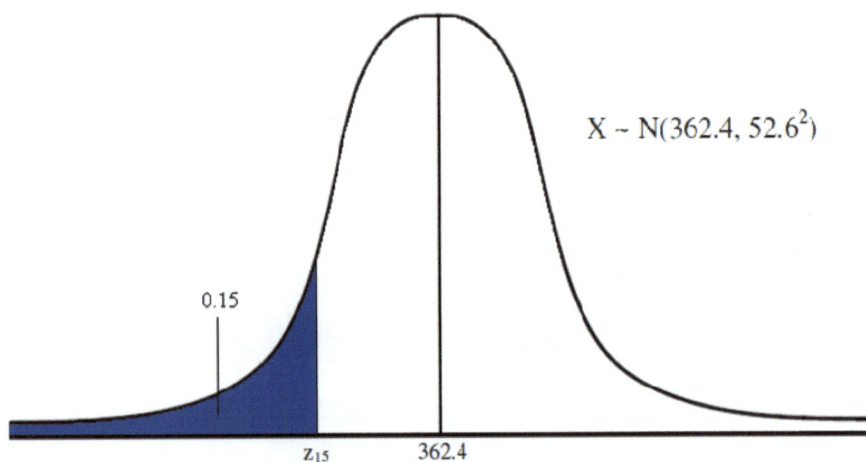

We need to find the standardized value z_{15}, below which 15% of the data values on the normal distribution lie. We therefore need to **de-standardize**, since we already have the value of the probability in the region under consideration. We therefore need to go within the body of the z-able to find the z-value that corresponds to the area under consideration represented in the graph.

At this point, you would have already gone through the process of de-standardizing, so that will not be repeated here. We therefore go directly to the answer.

Using any version of the z-table, we get that the standardized z-score that corresponds to the shaded area in the graph is -1.04.

Therefore, $z_{15} = -1.04$

We know that: $z_{15} = \dfrac{M_{15} - 362.4}{56.2}$

Therefore: $-1.04 = \dfrac{M_{15} - 362.4}{56.2}$

We now use this equation to find M_{15}.

Cross-multiplying: $(-1.04)(56.2) = M_{15} - 362.4$

$$-58.45 = M_{15} - 362.4$$

So, $M_{15} = 362.4 - 58.45$

$$= 303.95$$

Therefore the level of weekly milk production below which the cows will be sent to be meat-producers is 303.95 pounds. In other words, cows that produce less than 303.95 pounds of milk per week will be converted into beef!

Exercise 3.14

(1) The amount of time spent on a popular social networking site each day by computer science students at the University of Trinidad and Tobago is normally distributed with a mean of 2.6 hours and a standard deviation of 0.5 hours. Determine the proportion of students who spend more than 3 hours on the social networking site.

(2) On a coconut estate in Guyana, a 'good' coconut is considered to be one that produces between 0.25 and 0.34 litres of coconut water. If the amount of water produced by a coconut from this estate is normally distributed with a mean of 0.29 litres and a variance of 0.01 litres, determine the approximate percentage of 'good' coconuts on this particular estate.

(3) The data at the local telephone company shows that the average amount of time that their customers spend on the phone for each call is normally distributed with a mean of 2.77 minutes and a standard deviation of 1.01 minutes. The company wants to design a promotion aimed at luring people to spend more than 3 minutes on the phone. If a customer is selected at random from this company's database, what is the probability that that customer already spends an average of more than 3 minutes on the phone?

(4) Historical records at the Queen Elizabeth hospital in Barbados indicate that the average weight of a newborn female is 7.02 pounds. The standard deviation of the weights of newborn females at this hospital is 1.34 pounds. Using these statistics as a guide, approximate the number of newborn females out of the next one hundred that would have a newborn weight of between 6.78 and 7.45 pounds.

(5) The amount of time elapsed before the first goal is scored in a soccer game in a football league in England is normally distributed with a mean of 24.7 minutes and a standard deviation of 28.5 minutes. Based on these statistics, determine the approximate percentage of games in which the first goal is scored in the second half. *(Hint: A soccer game is comprised of two halves, each of 45 minutes duration)*

(6) At the University of the West Indies, a statistics lecturer decides to grade 'on the curve'. He decides that the top one percent of all grades will be awarded a distinction, while the bottom five percent of the grades will be given a 'fail' grade. If the marks for the exam are normally distributed with a mean 56 and a variance of 17, determine:
(i) the minimum grade for a student to be awarded a distinction
(ii) the 'pass mark' for this exam.

(7) An exclusive Port-of-Spain millionaires' club is besieged with applications from new millionaires seeking to join. The club's executive decides that of the applicants, only those whose earnings for the last year are in the top three percent will be accepted. The earnings for the last year of the applicants are normally distributed with a mean of $2,437,098 and a standard deviation of $234,651. Determine the minimum value of the last year's earnings below which an applicant will not be accepted into the millionaires club.

(8) International boxing rules state that the minimum weight that would qualify a boxer to fight as a heavyweight is 200 pounds. A boxing gym in Marabella is recruiting boxers for its heavyweight camp. Aspiring boxers are weighed and their weights are found to be normally distributed with a mean of 187.13 pounds and a standard deviation of 9.67 pounds. The management of the club decides to accept only those boxers whose weights fall within the upper seven percent of the weights of the aspiring boxers. Determine the minimum weight of the boxers who are accepted.

(9) Customer records of a mobile phone company in Uganda show that the amount of money spent per month by pre-paid customers buying phone cards is normally distributed with a mean of $477.84 and a standard deviation of $137.03. A telecommunications consultant has recommended that customers whose monthly expenditure on phone cards lies within the uppermost 10 percent be wooed into signing up for the post paid plan. She also recommends that certain promotions be directed at customers whose monthly expenditure falls within the lowest 10 percent, with a view to getting them to spend more money on phone cards. Customers whose monthly expenditure falls between these two values will be left alone. Calculate the two values of monthly phone card expenditure for which customers whose monthly expenditure falls between these two values will be left alone.

(10) The heart rates of a group of aspiring special forces commandos are measured and found to be normally distributed with a mean of 50.32 beats per minute and a standard deviation of 6.59 beats per minute. Only those commandos whose heart rates fall within the lowest 17 percent will be accepted. Determine then, the maximum heart rate of the accepted commandos.

4
Combining Normal Variables

Sometimes it is useful and necessary to combine two separate normal variables by adding them to each other, or by subtracting one from the other. Sometimes we may want to define a new normal variable which is simply a multiple of a prior normal variable. Sometimes we may wish to combine by addition or subtraction two variables that are each themselves multiples of another variable.

Multiples of a Normal Variable

Consider a normal variable X defined with a mean μ and a standard deviation σ (variance σ^2). Hence, $X \sim N(\mu, \sigma^2)$.

Suppose that we want to define a new normal variable nX, which is a multiple of the variable X.

The mean of the new normal distribution nX would be found as follows:
$$E(nX) = nE(X)$$
$$= n\mu$$

The variance of the new normal distribution nX would be found as follows:
$$Var(nX) = n^2 Var(X)$$
$$= n^2 \sigma^2$$

The new variable nX would thus have a normal distribution with a mean of '$n\mu$', and a variance of '$n^2\sigma^2$'. So, $nX \sim N(n\mu, n^2\sigma^2)$.

For $X \sim N(\mu, \sigma^2)$:
$$E(nX) = nE(X),$$
$$Var(nX) = n^2 Var(X).$$
Hence, $nX \sim N(n\mu, n^2\sigma^2)$

Example 4.1
If $X \sim N(5,3^2)$, define the variables 4X, and 2.5X.
$$E(4X) = 4E(X)$$
$$= 4(5)$$
$$= 20$$
$$Var(4X) = 4^2 Var(X)$$

$$= 16\text{Var}(X)$$
$$= 16(3^2)$$
$$= 16(9)$$
$$= 144$$

Therefore $4X \sim N(20, 144)$

(b) $E(2.5X) = 2.5E(X)$
$$= 2.5(5)$$
$$= 12.5$$
$$\text{Var}(2.5X) = (2.5)^2\text{Var}(X)$$
$$= 6.25\text{Var}(X)$$
$$= 6.25(3^2)$$
$$= 6.25(9)$$
$$= 56.25$$

Therefore, $2.5X \sim N(12.5, 56.25)$

The Sum of Two Independent Normal Variables

Consider a normal variable defined as $X \sim N(\mu_x, \sigma_x^2)$. Consider another normal variable defined as $Y \sim N(\mu_y, \sigma_y^2)$. Assume that these two distributions are independent of each other. Suppose that we wish to define a new random variable defined as the sum of the two normal variables X and Y. The sum of X and Y can be defined in one of two ways: 'X+Y' or 'Y+X'. Because we are adding the variables, it makes no difference whether we define the new variable as 'X+Y' or 'Y+X'. The final values of the parameters would be the same in each case.

The mean of the new normal distribution (X+Y) would be found as follows:
$$E(X+Y) = E(X) + E(Y)$$
$$= \mu_x + \mu_y$$
The variance of the new normal distribution (X+Y) would be found as follows:
$$\text{Var}(X+Y) = \text{Var}(X) + \text{Var}(Y)$$
$$= \sigma_x^2 + \sigma_y^2$$
The new variable (X+Y) would therefore have a normal distribution with mean of '$\mu_x + \mu_y$', and a variance of '$\sigma_x^2 + \sigma_y^2$'. Hence, $(X+Y) \sim N(\mu_x + \mu_y , \sigma_x^2 + \sigma_y^2)$

$$E(X+Y) = E(X) + E(Y)$$
$$E(Y+X) = E(Y) + E(X),$$
$$\text{and Var}(X+Y) = \text{Var}(X) + \text{Var}(Y)$$
$$\text{Var}(Y+X) = \text{Var}(Y) + \text{Var}(X)$$

$$E(Y+X) = E(X+Y)$$
$$\text{Var}(Y+X) = \text{Var}(X+Y)$$
The two normal distributions MUST be independent for these relationships to hold!

Example 4.2

If $X \sim N(9,4^2)$, and $Y \sim N(7,3^2)$, define a new random variable X+Y. Assume that the normal variables X and Y are independent of each other.

$$E(X+Y) = E(X) + E(Y)$$
$$= 9 + 7$$
$$= 16$$

$$Var(X+Y) = Var(X) + Var(Y)$$
$$= 4^2 + 3^2$$
$$= 16 + 9$$
$$= 25$$

Therefore, $(X+Y) \sim N(16,25)$

The Difference between Two Independent Normal Variables

Consider a normal variable defined as $X \sim N(\mu_x, \sigma_x^2)$. Consider another normal variable defined as $Y \sim N(\mu_y, \sigma_y^2)$. Assume that these two distributions are independent of each other. Suppose that we wish to define a new random variable defined as the difference between of the two normal variables X and Y. The difference between X and Y can be defined in one of two ways: 'X-Y' or 'Y-X'. The definition we use would depend on the precise nature of the situation before us, as well as our precise goal in resolving a situation involving the difference between two normal variables.

The mean of the new normal distribution (X+Y) would be found as follows:
$$E(X-Y) = E(X) - E(Y)$$
$$= \mu_x - \mu_y$$
The variance of the new normal distribution (X+Y) would be found as follows:
$$Var(X+Y) = Var(X) + Var(Y)$$
$$= \sigma_x^2 + \sigma_y^2$$
The new variable (X+Y) would therefore have a normal distribution with mean of '$\mu_x + \mu_y$', and a variance of '$\sigma_x^2 + \sigma_y^2$'. Hence, $(X-Y) \sim N(\mu_x - \mu_y , \sigma_x^2 + \sigma_y^2)$

E(X-Y) = E(X) – E(Y)
E(Y-X) = E(Y) – E(X),
and Var(X-Y) = Var(X) + Var(Y)
Var (Y-X) = Var(Y) + Var(X)

E(Y-X) = - E(X-Y)
Var(Y-X) = Var(X-Y)
The two normal distributions MUST be independent for these relationships to hold!

*Note that Var(X-Y) **IS NOT EQUAL TO** Var(X) – Var(Y)!*

Var(X-Y) = Var(X+Y) = Var(X) + Var(Y)
Independent variances are *ALWAYS* added to each other!

Example 4.3
Consider the same two normal variables X ~ $N(9,4^2)$, and Y ~ $N(7,3^2)$. Define the following new random variables: X-Y, and Y-X. Assume that the variables are independent of each other.

(a) E(X-Y) = E(X) – E(Y)
$$= 9 - 7$$
$$= 2$$

Var(X-Y) = Var(X) + Var(Y)
$$= 4^2 + 3^2$$
$$= 16 + 9$$
$$= 25$$
Therefore, (X-Y) ~ N(2, 25)

(b) E(Y-X) = E(Y) – E(X)
$$= 7 - 9$$
$$= -2$$

Var(Y-X) = Var(Y) + Var(X)
$$= 3^2 + 4^2$$
$$= 9 + 16$$
$$= 25$$
Therefore, (Y-X) ~ N(-2, 25)

A mean of -2 for the variable (Y-X) is nothing to be alarmed about. What this means is that Y is greater than X by an average value of -2, in other words, Y is *less* than X by an average value of 2. This is the same as saying that X is greater than Y by an average value of 2, which is exactly what is meant by E(X-Y) = 2. So while mathematically, E(Y-X) is equal to -E(X-Y), in practical real-world terms, they are saying precisely the same thing. The variable we select to be the 'greater' value when defining a random variable in terms of the difference between two independent normal variables depends on what exactly we are trying to determine.

Exercise 4.1

(1) If $X \sim N(105, 15^2)$, determine the following distributions:
(a) $2X$ **(b)** $3X$ **(c)** $2.3X$ **(d)** $4.06X$

(2) If $P \sim N(0.05, 0.1^2)$, determine the following distributions:
(a) $0.5P$ **(b)** $12P$ **(c)** $5.77P$ **(d)** $7.5P$

(3) If $K \sim N(3.4, 1.1^2)$, determine the following distributions:
(a) $14K$ **(b)** $6K$ **(c)** $8.21K$ **(d)** $0.024K$

(4) If $U \sim N(2349, 15^2)$, determine the following distributions:
(a) $2U$ **(b)** $0.77U$ **(c)** $0.04U$ **(d)** $120U$

(5) If $F \sim N(54, 4.5^2)$, determine the following distributions:
(a) $8N$ **(b)** $12.47N$ **(c)** $0.9N$ **(d)** $265N$

(6) If $X \sim N(78, 6^2)$, and $Y \sim N(66, 4^2)$, determine the following distributions:
(a) $X + Y$ **(b)** $X - Y$ **(c)** $2X + 2Y$ **(d)** $2X - 2Y$

(7) If $P \sim N(12, 2^2)$ and $Q \sim N(16, 3^2)$, determine the following distributions:
(a) $P + Q$ **(b)** $P - Q$ **(c)** $2P + Q$ **(d)** $P - 2Q$

(8) If $C \sim N(0.5, 0.1^2)$, and $D \sim N(0.73, 0.13^2)$, determine the following distributions:
(a) $C - D$ **(b)** $D - C$ **(c)** $2D - 4C$ **(d)** $3C - 2D$

(9) If $T \sim N(1500, 54^2)$, and $V \sim N(1765, 67^2)$, determine the following distributions:
(a) $V - T$ **(b)** $T - V$ **(c)** $2T + V$ **(d)** $2T - V$

(10) If $G \sim N(455, 25^2)$, and $H \sim N(303, 14^2)$, determine the following distributions:
(a) $G + H$ **(b)** $H + G$ **(c)** $G - H$ **(d)** $H - G$

Example 4.3
The average weight of persons entering a weight-loss clinic is normally distributed with a mean of 302.7 pounds, and a variance of 225.87 pounds. The average weight of persons leaving the clinic upon completion of their program is normally distributed with a mean of 207 pounds with a variance of 82.3 pounds. Based on this data alone, determine the approximate proportion of persons who would leave the clinic having lost less than 50 pounds.

Solution
Our first step as always in problems of this type, is to define the random variables.

Let X be the random variable 'weight of persons entering the weight-loss clinic'
$$X \sim N(302.7, 225.87)$$
Let Y be the random variable 'weight of persons leaving the weight-loss clinic'
$$Y \sim N(207, 82.3)$$

We are dealing here with a situation where people go to a clinic to *lose* weight, so what we are dealing with is a situation where we have a *difference* between the weight of a person entering the clinic, and the weight of a person leaving the clinic. So we define a new variable for the *difference* between the weight entering the clinic, and the weight leaving the clinic.

So, we let D = X - Y.

We now need to establish the parameters for our variable 'D'.

$$E(D) = E(X) - E(Y)$$
$$= 302.7 - 207$$
$$= 95.7$$

$$Var(D) = Var(X) + Var(Y)$$
$$= 225.87 + 82.3$$
$$= 308.17$$

Therefore, D ~ N(95.7, 308.17)

We now need to determine the probability that a random patient will leave the clinic having lost less than 50 pounds. In other words, we are finding the probability that the *difference* between the patient's weight entering the clinic and their weight leaving the clinic is less than 50 pounds. Graphically, we have:

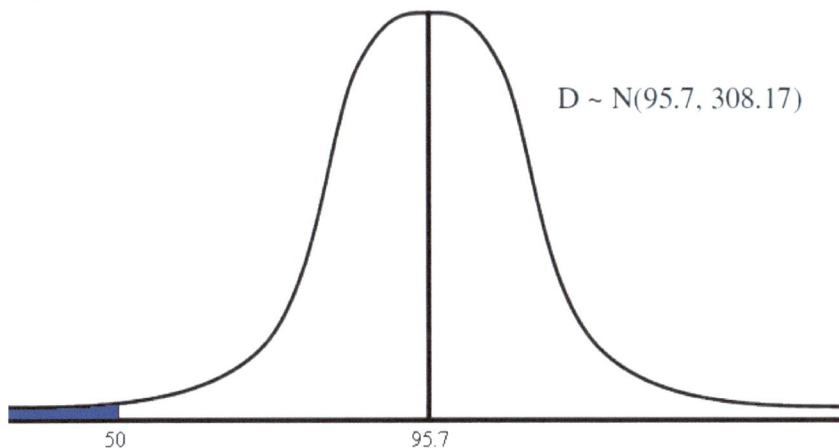

D ~ N(95.7, 308.17)

We are looking for P(D < 50)

We must now standardize to find the z-value that represents '50' on this particular distribution.

$$P(D < 50) = P(z < \frac{50 - 95.7}{\sqrt{308.17}})$$
$$= P(z < \frac{50 - 95.7}{17.56})$$
$$= P(z < -2.60)$$
$$= 0.0047$$

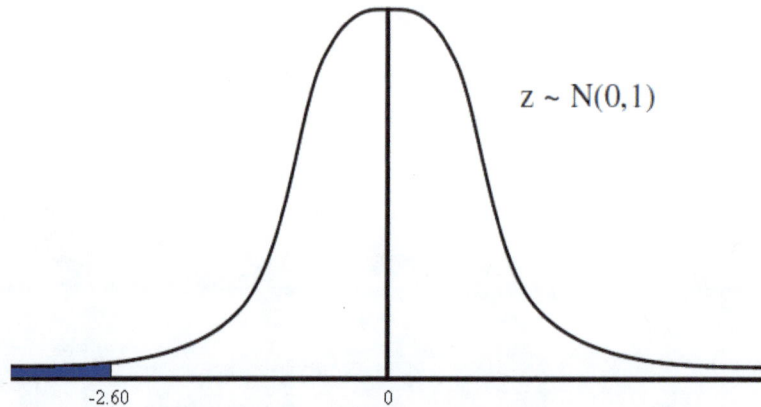

$z \sim N(0,1)$

-2.60   0

To express this probability as a proportion, we multiply by 100.
0.0047 x 100 = 0.47

Therefore, 0.47 percent of the patients leaving this clinic will leave having lost less than 50 pounds. What this means is that 0.47 out of every 100 patients (4.7 per 1,000 or 47 of every 10,000) entering the clinic will leave less than 50 pounds lighter. What a record!

Example 4.4
At a school bazaar, Chris decides to offer a randomly selected boy and a randomly selected girl a lift home. However, he is being careful not to overload his car, and he has the requisite instruments that will alert him as to whether the car is overloaded or not. An overload occurs if the combined weight of the boy and the girl selected exceeds 200 pounds. The weights of the boys at the bazaar are normally distributed with a mean of 122.3 pounds and a standard deviation of 24.5 pounds. The girls' weights are normally distributed with a mean of 105 pounds and a standard deviation of 16.3 pounds. How likely is it that Chris' car would be overloaded at the first selection of a boy and a girl?

Solution
As always, we begin by defining the random variables.

Let B be the random variable 'weights of boys in the bazaar'
$$B \sim N(122.3, 24.5^2)$$

Let G be the random variable 'weights of girls in the bazaar'
$$G \sim N(105, 16.3^2)$$

Our situation here is one where we are considering the *sum* of boys' weights and girls' weights. So we define a new variable for the *sum* of the boys' and the girls' weights.
So, we let S = B + G.
We now need to establish the parameters for our variable 'S'.

E(S) = E(B) + E(G)
      = 122.3 + 105
      = 227.3

Var(S) = Var(B) + Var(G)
       = $24.5^2 + 16.3^2$
       = 600.25 + 265.69
       = 865.94

Therefore, S ~ N(227.3, 865.94)
So, we now look for P(S > 200).
Graphically, we have,

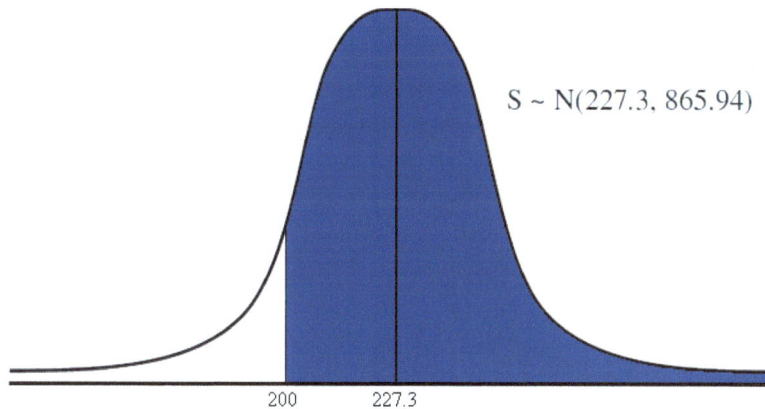

S ~ N(227.3, 865.94)

200   227.3

We must now standardize to find the z-value that represents '200' on this particular distribution.

$$P(S > 200) = P(z > \frac{200 - 227.3}{\sqrt{865.94}})$$

$$= P(z > \frac{200 - 227.3}{29.43})$$

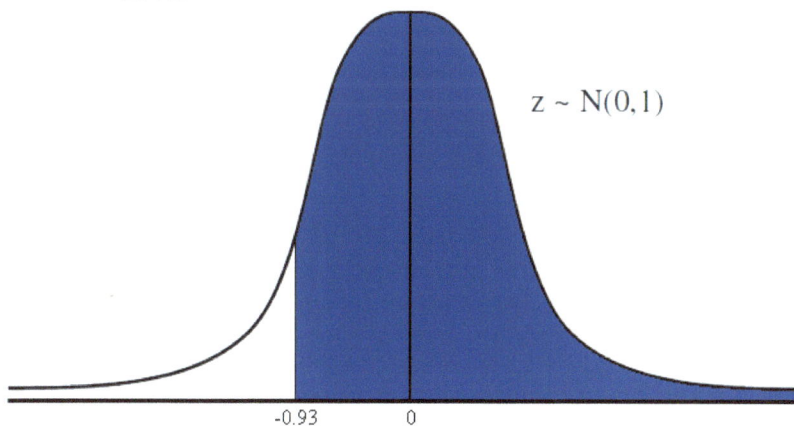

z ~ N(0,1)

-0.93   0

= P(z > - 0.93)
= 0.7357

Therefore there is a likelihood of 0.7357 that Chris' car would be overloaded on the first selection.

Exercise 4.2

(1) At a certain high school in South Carolina, USA, an aptitude test is taken upon entry. This test consists of a Mathematics examination and an English examination. The overall score on the test is computed by adding the Math score and the English score, both exams being scored out of 900. Data from past tests show that the Math score is normally distributed with a mean of 586 and a standard deviation of 97. The English score is normally distributed with a mean of 532 and a standard deviation of 86. Using this data as a guide, find:

(a) the proportion of applicants from the latest test who can be expected to have a score combined score greater than 1050.

(b) the proportion of students who can be expected to have a combined score of between 700 and 900.

(c) the minimum score for which a student will be awarded a distinction, given that distinctions are awarded to students with the highest 10% of the scores.

(2) A certain technically oriented high school administers the same test to prospective students. The overall score is found by computing a weighted sum, where three-fifths of the Math score is added to two-fifths of the English score. Find the probability that a randomly selected student scores between 650 and 750 marks on the test.

(3) An artistically oriented high school administers the same test to its prospective students. In this case, the overall score is found by computing a weighted sum where two-fifths of the Math score is added to three-fifths of the English score. Determine the likelihood of a student scoring between 250 and 350 marks on the test.

(4) A weight loss clinic in Westmoorings, Trinidad and Tobago maintains records of the weights of its patients before and after treatment. The weights of the past patients before treatment are normally distributed with a mean of 267 pounds and a standard deviation of 12 pounds. The weights of the patients after treatment are normally distributed with a mean of 192 pounds and a standard deviation of 19 pounds. A curious salesman for the clinic decides to pull the records of a randomly selected past patient to demonstrate the efficacy of the clinic's treatment. What is the probability that the difference between the 'before' and 'after' weights of this patient is greater than 50 pounds?

(5) Two competing banks in Mumbai, India decide to measure the time spent by each customer waiting in lines to be served. At the first bank, the time spent waiting in line is normally distributed with a mean of 35.7 minutes and a standard deviation of 15.3 minutes. At the second bank, the time spent waiting in line is normally distributed with a mean of 53 minutes and a standard deviation of 11 minutes. A customer has taken the morning off from her job to attend to her affairs in both banks. What is the probability that, between the two banks, she will spend more than two hours waiting in lines?

(6) There are two direct routes between Arima and Port-of-Spain - the Priority Bus Route, and the Eastern Main Road. The Priority Bus Route is used exclusively by buses, while all other traffic uses the Eastern Main Road. The time taken for any vehicle to travel from Arima to Port-of-Spain on the Priority Bus Route is normally distributed with a mean of 29.6 minutes and a standard deviation of 12 minutes. The time taken for any vehicle to travel from Arima to Port-of-Spain on the Eastern Main Road is normally distributed with a mean of 49.8 minutes and a standard deviation of 8 minutes. If one vehicle on each route is randomly selected and timed leaving Arima at the same time heading into Port-of-Spain, find the probability that the vehicle travelling on the the Priority Bus Route will reach Port of Spain more than 15 minutes before the vehicle travelling on the Eastern Main Road.

(7) The time taken by Barber Smith to do a single haircut is normally distributed with a mean of 21.4 minutes and a standard deviation of 6.8 minutes. Joe visits the barber and finds that there are two people ahead of him in line to get haircuts. Find the probability that Joe would start his haircut within half-an-hour.

(8) A certain gymnastics school mandates that drills performed by mixed-sex couples involve only couples where the woman is at least 60 pounds lighter than the man. The weights of the men in this school are normally distributed with a mean of 185.9 pounds and a standard deviation of 13.4 pounds. The weights of the women are normally distributed with a mean of 118 pounds with a standard deviation of 14.4 pounds. Suppose that a man and a woman are selected randomly to perform a drill. What is the probability that they would be suitable for each other according to the rules of the school?

(9) The see-saw at the local park is designed to go off-balance if the difference between the weights of the persons at both ends exceeds 50 pounds. The weights of boys who play in the park are normally distributed with a mean of 101.3 pounds and a standard deviation of 9.1 pounds. The weights of girls who play in the park are normally distributed with a mean 81 pounds, with a standard deviation of 8.3 pounds. Find the probability that the see-saw would go off-balance if:
(a) one boy and one girl sit at opposite vehicle to ends of the see-saw.
(b) four boys sit at one end, and three girls sit the other end of the see-saw.
(c) three boys sit at one end, and four girls sit at the other end of the see-saw

(10) A chicken farm in Rousillac in South Trinidad is launching a massive marketing campaign to compete with a similar chicken farm in Mausica in East Trinidad. Their sales representatives claim that their chickens are so superior to those of their competitors, that three Rousillac chickens weigh more than four Mausica chickens. If the weights of the Rousillac chickens are normally distributed with a mean of 5.7 pounds and a standard deviation of 0.82 pounds, and the weights of the Mausica chickens are normally distributed with a weight of 5.3 pounds and a standard deviation of 1.02 pounds, find the probability that three random Rousillac chickens and four random Mausica chickens, when weighed, will yield results that support the claim of the Rousillac sales representatives.

5

The Sampling Distributions

The sampling distributions form an integral part of statistics practice. Theoretically, a sampling distribution is a distribution formed from all the possible *samples* of a particular size taken from a specified population. Practically however, we need only one sample to construct a working sampling distribution. The two sampling distributions we will treat with are the *Distribution of the Sample Mean* and the *Distribution of the Sample Proportion*.

THE DISTRIBUTION OF THE SAMPLE MEAN

The development of the Distribution of the Sample Mean is governed by the interaction of three factors – (a) The nature of the population
(b) The size of the sample
(c) The availability of population parameters

Normal Populations

Suppose that we have a specific population consisting of a class of 100 university students. We know that the amount of money that a randomly chosen student from this population spends every week buying lunch from Top Lunch Café is normally distributed. We would find the mean amount of money spent on lunch by the students of this population by summing the amount spent on lunch by each of the 100 students and then dividing by 100. Using this mean, we would employ the process used in chapter 1 to find the standard deviation of the amount of money spent on lunch by the students of this class.

Suppose instead that we wanted to find the mean amount of money spent on lunch by a *randomly* chosen sample of 40 students from this class of 100? We would need to select 40 students randomly, sum the amount spent on lunch by each of those 40 students, and then divide this sum by 40. As in the case of the entire population, we then use this mean to calculate the standard deviation of the amount of money spent on lunch by students from this random sample of 40 students, using the procedure outlined in our first chapter. The problem here is that our randomly chosen sample of size 40 is but one of literally *millions* of possible samples of size 40 out of a population of size 100. To be precise, the number of samples of size 40 which can be drawn from a population of size 100 is equal to: 13,746,234,150,000,000,000,000,000,000!!

So theoretically, we would have to find the means of all 13.75 octillion samples of size 40, and then divide by 13.75 octillion. Try to imagine what would happen if the population size was 1,000, or better yet, 10,000! Practically speaking, we have neither the time nor the inclination to do this many calculations, so we take one sample of size 40, and use the parameters from this one sample to construct a working distribution of the sample mean.

If we know the value of the population mean and the population variance for a normal population, statistics theory tells us that the distribution of the sample mean is normal with a mean equal to the population mean, with a variance that is found by dividing the population variance by the sample size.

The Distribution of the Sample Mean is the probability distribution of the means of all possible samples of a specified size from a normal population. The probability distribution lists each possible value of the sample mean and the probability of obtaining that value when we measure the mean of a randomly selected sample of a particular size.

The probability distribution of the sample mean is normal with a mean equal to the population mean μ, and a variance equal to $\dfrac{\sigma^2}{n}$ (standard deviation $\dfrac{\sigma}{\sqrt{n}}$).

So, if $X \sim N(\mu, \sigma^2)$, then $\overline{X} \sim N(\mu, \dfrac{\sigma^2}{n})$ for samples of size n.

Be very careful in noting the difference between the random variable 'X' and the random variable '$\overline{X}$'. 'X' is the probability distribution of **all possible values** of a particular variable **for all members of a population**, while '$\overline{X}$' is the probability distribution of the **all possible values of the sample mean for all possible samples of size n** drawn from that population.

Example 5.1
The heights of the boys in a particular high school are normally distributed with a mean of 160 cm and a standard deviation of 9.2 cm . Find the probability that a boy from a random sample of 55 boys has a height greater than 163 cm.

Solution
The first thing we note here is that we are dealing with a *sample*, and not a population. We are dealing with a boy selected from a random sample of 55, and not a boy selected from the entire population of the high school. Therefore we use the distribution of the sample mean to treat with this situation. Our first step then, is to define the random variable:

Let $\overline{H}$ be the random variable 'height of a boy from a random sample of size 55'.

$$\overline{H} \sim N(160, \frac{9.2^2}{55})$$
$$\therefore \overline{H} \sim N(163, 1.539)$$

We are required to find: P($\overline{H}$ > 163):

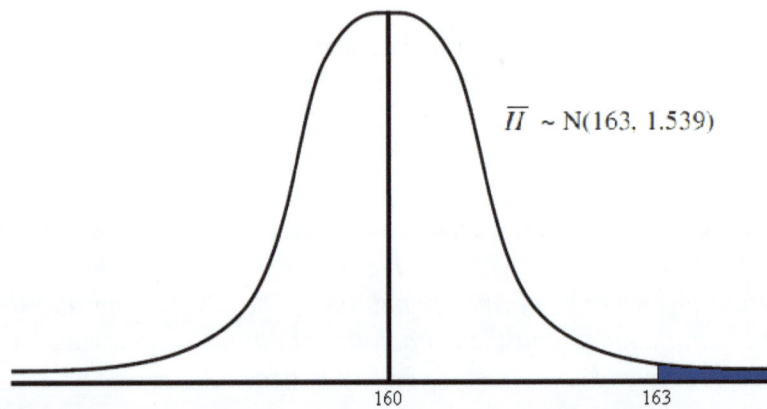

$$\overline{H} \sim N(163, 1.539)$$

We need to 'standardize' the value 163, so that we can use the z – table to calculate the required probability:

$$z_{163} = \frac{163 - 160}{\sqrt{1.539}}$$

$$= 2.42$$

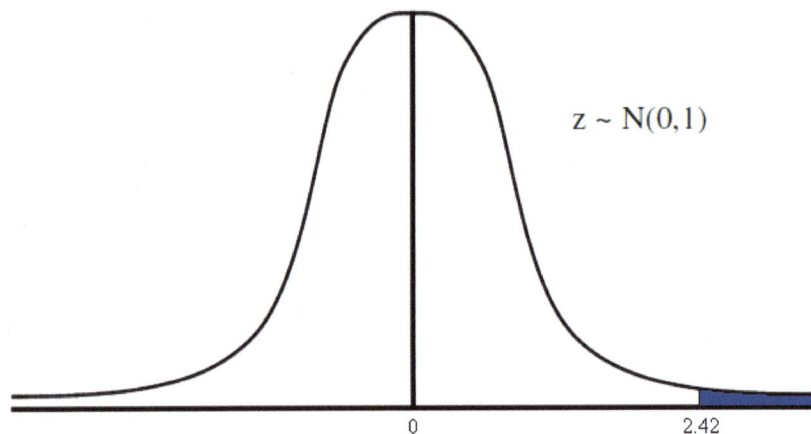

$$z \sim N(0,1)$$

Therefore, $P(\overline{H} > 163) = P(z > 2.42)$
$$= 0.0078$$

Therefore, the probability that a boy from a random sample of 95 boys has a height greater than 163 cm is 0.0078.

Large Non-Normal or Unknown Populations

If the population is known to be non-normal, or its distribution is unknown, we would then need to consider the size of the sample in formulating the Distribution of the Sample Mean. If the sample is 'large', the distribution of the sample mean is formed in the same manner as if the population was normal. If the sample is small, we cannot use the Normal Distribution to form the distribution of the sample mean. We would then have to use the Student's t-Distribution.

As far as non-normal or unknown populations are concerned, we shall restrict our consideration only to the case of large samples. In terms of distinguishing between a small sample and a large sample, take note of and remember the following:

A 'large' sample is taken as one with a sample size of 30 or more. (n ≥ 30).
A 'small' sample is taken as one with a sample size of less than 30. (n < 30).

The theoretical foundation that allows us to obtain a distribution of the sample mean that is normally distributed when a large sample is drawn from a non-normal or unknown population is the **Central Limit Theorem**.

The Central Limit Theorem for Sample Means

For a large sample of size 'n', the distribution of the sample mean is normal, whether or not the population itself is normal.

When the population mean μ is known, the mean of the distribution of the sample mean is μ. When the population mean is unknown, the sample mean $\bar{x}$ is used as a point estimate of the population mean, and the mean of the distribution of the sample mean is $\bar{x}$.

When the population variance σ^2 is known, the variance of the distribution of the sample mean is $\dfrac{\sigma^2}{n}$ (standard deviation $\dfrac{\sigma}{\sqrt{n}}$). When the population variance is unknown, the sample variance s^2 is used as a point estimate of the population variance, and the variance of the distribution of the sample mean is $\dfrac{s^2}{n}$ (standard deviation $\dfrac{s}{\sqrt{n}}$).

What this means is that for large populations the distribution of the sample mean *behaves normally,* regardless of the nature of the population itself.

Availability of Parameters

Even if we know for a fact that a given population is normal, we may not have both population parameters available for construction of the distribution of the sample mean. In that case we have to lean on the sample statistics. When we know the population mean μ and the population variance σ^2, we use these straightaway to derive the parameters for the distribution of the sample mean.

Whenever the population mean or the population variance are unknown, we use the sample mean $\bar{x}$ and sample variance s^2 respectively as *point estimators* of the population mean and population variance respectively. The distribution of the sample mean is then constructed accordingly.

When the sample is large, even if the population is not normal, we can construct the distribution of the sample mean in an identical manner to the one prescribed above for normal populations, taking the availability or non-availability of population parameters into account. The table below summarizes this information. See the beginning of chapter 6 for a more detailed explanation of point estimators.

Population Variance σ^2

	KNOWN	UNKNOWN
KNOWN	$\overline{X} \sim N(\mu, \dfrac{\sigma^2}{n})$	$\overline{X} \sim N(\mu, \dfrac{s^2}{n})$
UNKNOWN	$\overline{X} \sim N(\overline{x}, \dfrac{\sigma^2}{n})^{*}$	$\overline{X} \sim N(\overline{x}, \dfrac{s^2}{n})^{*}$

Population Mean μ (row label)

* These are the distributions that are used to construct confidence intervals for the unknown population mean μ.

Exercise 5.1
Construct the distribution of the sample mean in each of the following scenarios:

(1) $X \sim N(151,15^2)$
(a) sample size = 45   **(b)** sample size = 26   **(c)** sample size = 64   **(d)** sample size = 17

(2) $P \sim N(0.3, 0.2^2)$
a) sample size = 29   **(b)** sample size = 205   **(c)** sample size = 11   **(d)** sample size = 99

(3) A sample of size 63 from a population whose mean and variance are unknown is found to have a mean of 132.56 and a standard deviation of 9.43

(4) A sample of size 22 from a normal population whose mean is 56.3 and whose variance is unknown is found to have a variance of 11.

(5) A sample of size 96 from a population whose mean is 71.11 and whose variance is unknown is found to have a standard deviation of 8.64.

(6) A sample of 40 footballs from a factory in Pakistan is found to have a mean diameter of 25.4 cm, with a standard deviation of 0.5 cm.

(7) A sample of 39 cement bags at a factory in Mexico is found to have a mean weight of 80 lbs with a variance of 2.67 lbs.

(8) The average time spent sleeping each night among a sample of 50 final year university students is found to be 5.66 hours. The variance for this sample is 1.8 hours.

70

(9) The concentration of cement in a sample of 65 bags of concrete mix is found to have a mean of 0.12, with a standard deviation of 0.02

(10) The concentration of alcohol in a random sample of 100 beer bottles at a beer distillery is found to have a mean of 0.063 with a standard deviation of 0.014.

Exercise 5.2

(1) The weights of newborn babies in a hospital are known to be normally distributed with a mean of 7.2 pounds and a variance of 2 ounces. What percentage of a random sample of 50 newborn babies from this hospital can be statistically expected to weigh more than 7.6 pounds?

(2) The scores recorded by divers in the preliminaries of an Olympic competition are normally distributed with mean 8.905 and standard deviation 1.002. How likely is it that a diver from a random sample of 30 divers from the preliminaries would have a score less than 8.867?

(3) The earthquake measurements in the town of San Francisco for a period of one year are measured on the Richter scale. These measurements are normally distributed with a mean 5.33 and variance 2.25. From these records, a sample of 35 earthquakes is selected for analysis. What is the probability that a randomly selected record from this sample has a Richter scale measurement of more than 5.75?

(4) The daily temperatures in degrees Celsius for the year 2008 in Trinidad and Tobago are normally distributed with mean 31.4 and standard deviation 1.23. Sixty-five days in 2008 are randomly selected. From this sample, what is the probability that the temperature is between 31.6 and 31.8 degrees Celsius?

(5) The heart rates of all the clients in Dr. Thomas' office are known to be normally distributed with mean 65 beats per minute and standard deviation 12 beats per minute. From a random sample of 40 of Dr. Thomas' patients, find the probability that a patient would have a heart rate below 59 beats per minute. Discuss your result.

(6) The local representative of the National Chemical Agency selects a sample of 44 specimens of a particular hair growth product for analysis of the minoxidil content of the specimens. The proportion of minoxidil in the specimens of this sample has a mean of 0.05 and a standard deviation of 0.0125. What is the probability that a specimen from this sample has a proportion of minoxidil that is higher than 0.051?

(7) The personal best-times in seconds for the 100 meter dash of a sample of 37 world-class sprinters has a mean of 10.02 and standard deviation of 0.07. What is the probability that a randomly selected sprinter from this sample would have a 100 meter best-time of between 9.99 and 10.05?

(8) The average distance travelled by the putts of a sample of 107 golfers at the local Country club is 203.45 meters. The variance of the distances travelled by these putts is 37 meters. What proportion of the putts from this sample can be expected to travel a distance greater than 204 meters?

(9) The mean amount of time spent daily looking at television by a sample of 33 residents of Televille is 65 minutes, with a standard deviation of 14 minutes. What is the likelihood that someone from this sample spends between 62 minutes and 67 minutes looking at television?

(10) The exam scores (out of 100) for a sample of 40 students from a large university class has a mean of 47 and standard deviation 23. From this sample, what is the probability that a randomly selected student would have a score greater than 50?

THE DISTRIBUTION OF THE SAMPLE PROPORTION
Large Samples

Suppose we want to find the proportion of the population of 100 university students that buy food at Top Lunch Cafe. We simply count the number of students from this population and divide this number by 100. Suppose however, that we want to find the proportion of a random sample of 40 of these students who buy food at Top Lunch Cafe. From this sample, we count the number of students who buy food at Top Lunch Cafe and divide by 40. If we wish to find the mean proportion of students from randomly chosen samples of size 40 who buy food at Top Lunch Cafe, then we would need to find the proportions for each of the 13,746,234,150,000,000,000,000,000,000 possible samples of size 40, add them all, and divide by 13,746,234,150,000,000,000,000,000,000! As in the case of the distribution of the sample mean, we do not have time or the inclination for all this calculation, and we only need one sample to construct a working distribution of the sample proportion. Let us consider the scenario in a bit more detail:

If we know for a fact that 80 of the 100 fast-food buying university students buy food at Top Lunch Cafe, then we can say that the proportion p of fast-food buying students who buy food at Top Lunch Cafe is 0.80 ($\frac{80}{100}$). This is an example of a population proportion, which is simply the proportion of a population that possesses a given characteristic. We can use the knowledge of the population proportion to construct the distribution of the sample proportion for samples of size n. Let us for example take a sample of size 40 from this population of 100 fast-food buying students, 80 of whom are known to buy food at Top Lunch Cafe.

If we take all 13.75 octillion possible samples of size 40 from the population of 100 and in each sample, we measure the proportion of fast-food eating students who patronize Top Lunch Cafe, we would find that these sample proportions are normally distributed with a mean equal to the population p, and a variance equal to $\frac{pq}{n}$ (standard deviation $\sqrt{\frac{pq}{n}}$) where $q = 1 - p$.

So when the population proportion is known to be 0.80, then the mean of the distribution of the sample proportion for samples of size 40 would be 0.80, and the variance would be $\frac{(0.80)(0.20)}{40} = 0.004$.

If we know the proportion p of a population that possesses a particular characteristic, then, for a sample of size n, the distribution of the sample proportion would have a mean of p and a variance of $\frac{pq}{n}$. If, however, the proportion p for the population is unknown, then, provided the sample is sufficiently large, the distribution of the sample mean would have a mean of $\hat{p}$, and a variance of. $\frac{\hat{p}\hat{q}}{n}$. In the case of the distribution of the sample proportion, a sample is deemed to be 'large' if $n\hat{p} > 5$, and $n\hat{q} > 5$. Once these conditions are satisfied, the sample proportion is used as a point estimator for the population proportion, and the distribution is formed accordingly. The following table provides a summary:

Population Proportion 'p'

KNOWN	UNKNOWN
$\hat{p} \sim N(p, \frac{pq}{n})$	$\hat{p} \sim N(\hat{p}, \frac{\hat{p}\hat{q}}{n})*$

* This is the distribution that is used to construct confidence intervals for the unknown population proportion p (see chapter 6).

As is the case with the distribution of the sample mean, it is the Central Limit Theorem that facilitates the construction of the distribution of the sample proportion for large populations when the population parameters are unavailable.

The Central Limit Theorem for Sample Proportions

For a known population proportion p, the distribution of the sample proportion is normal, with a mean of p, and a variance of $\frac{pq}{n}$ (standard deviation $\sqrt{\frac{pq}{n}}$).

For an unknown population proportion p, if the size of the sample is sufficiently large, the distribution of the sample proportion is normal, with a mean equal to the sample proportion $\hat{p}$ and variance of $\frac{\hat{p}\hat{q}}{n}$ (standard deviation of $\sqrt{\frac{\hat{p}\hat{q}}{n}}$).

A sample is considered to be sufficiently large if $n\hat{p} > 5$, and $n\hat{q} > 5$,

where: n = sample size, and $\hat{q} = 1 - \hat{p}$

Exercise 5.3

Construct the distribution of the sample proportion in the following scenarios:

(1) Half of a population possesses a particular characteristic, and samples of the following sizes are drawn from that population:

(a) 30   **(b)** 60   **(c)** 99   **(d)** 145

(2) 49 percent of a population possesses a particular characteristic, and samples of the following sizes are drawn from that population:

(a) 61   **(b)** 235   **(c)** 115   **(d)** 49

(3) Four out of every five Puerto Ricans are bilingual, and 40 Puerto Ricans are randomly sampled and surveyed.

(4) At a certain hospital, 5% of newborn babies are born on their 'due date'. The records of 200 random babies from this hospital are selected for analysis of their 'due date'.

(5) A random sample of 39 credit card customers of a certain bank are chosen for a survey on the use of their credit cards. 60% of the bank's customers who use credit cards use Mastercard.

(6) From samples of size 100, the following proportions possess a particular characteristic:

(a) 0.25   **(b)** 0.63   **(c)** 0.88   **(d)** 0.17

(7) From samples of the following sizes, 63 possess a particular characteristic:

(a) 126   **(b)** 70   **(c)** 100   **(d)** 243

(8) 21 out of a sample of 72 inmates from a particular prison are left-handed.

(9) 19 percent of a sample of 100 cows on a farm is found to be infected with mad cow disease.

(10) 13 of a random sample of 860 new manufactured smart phones contain defective software.

Example 5.2
Problem
Eighty percent of the mobile phone users in a particular town are under the age of thirty. If 100 mobile phone users from this town are randomly selected, find the probability that more than 90 percent of the sample are under the age of 30.
Solution
Okay, we immediately recognize that the population proportion p is 0.80 (eighty percent). We have a sample of size 100.

$$\therefore \ n = 100, \ p = 0.80, \ q = 0.20$$
$$np = (100)(0.80) = 80 : nq = (100)(0.20) = 20$$

Since both np and nq are greater than 5, we are now free to use the distribution of the sample proportion.

Our next step is to define a random variable for the distribution of the sample proportion in this instance. Our variable is defined as follows:

Let $\hat{p}$ be the random variable 'proportion of mobile phone users under the age of 30 from a random sample of 100 mobile phone users'.

Therefore: $\sim N(0.80, \dfrac{(0.80)(0.20)}{100})$

We are required to find $P(\hat{p} > 0.9)$. We are actually trying to determine the probability that the *proportion* of mobile phone users in the sample who are under 30 years old is greater than 0.9.

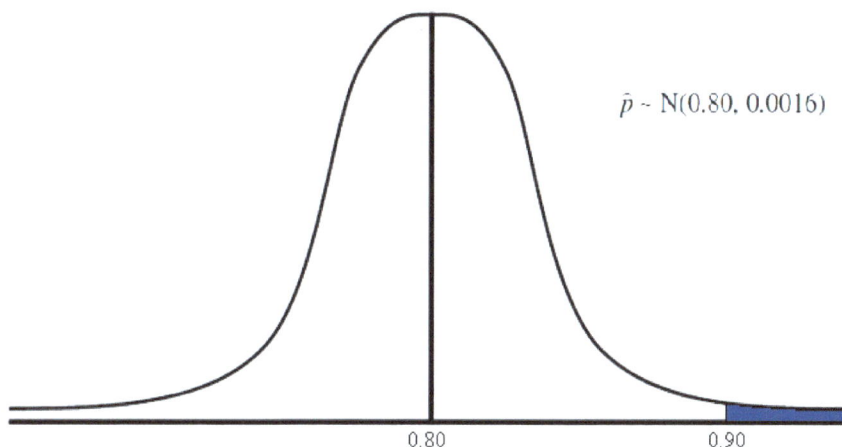

$\hat{p} \sim N(0.80, 0.0016)$

So, standardizing: $P(\hat{p} > 0.9) = P(z > \dfrac{0.9 - 0.8}{\sqrt{0.0016}})$

$$= P(z > \dfrac{0.9 - 0.8}{0.04})$$

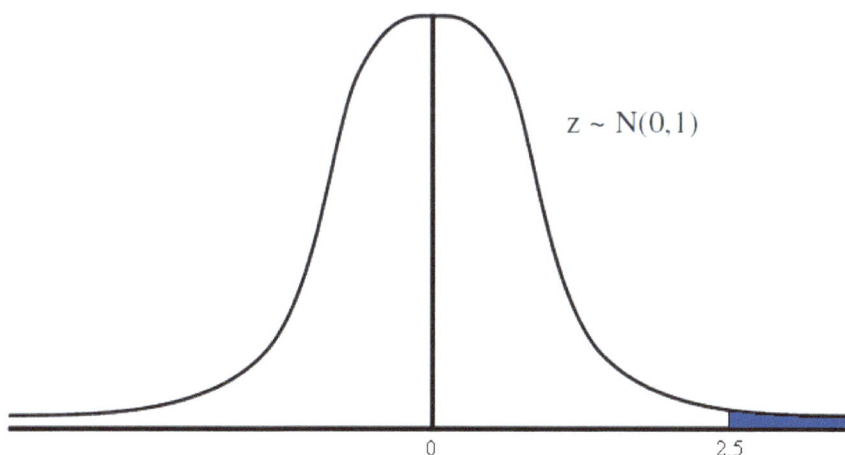

$z \sim N(0,1)$

$$= P(z > 2.5)$$
$$= 0.0062$$

Therefore the probability that more than 90 percent of the sample are under 30 years old = 0.0062

Note that it is not necessary to know the size of the population in order to develop the distribution of the sample proportion. In fact, if the size of the population is known, the formula for the variance/standard deviation would be slightly different to the one here. We will therefore keep things simple and leave that situation outside of our present discussion. Generally speaking, as is the case with the distribution of the sample mean, most problems and situations involving the distribution of the sample proportion involve populations that are substantially larger than the sample, and whose exact number is usually quite unknown.

Exercise 5.4

(1) 15% of drivers pulled over by the police last year for traffic violations were driving using expired driving permits. Assume that this figure is true for all drivers at present. Find the probability that of the next 45 drivers pulled over by the police, more than 17% are using expired driving permits.

(2) If four out of every five Puerto Ricans are bilingual, determine the probability that between seventy and ninety percent of a random sample of 60 Puerto Ricans would be bilingual.

(3) Unofficial estimates suggest that only seven percent of religious e-mails are forwarded at least once. If 99 such e-mails are chosen are random for analysis, find the probability that more than ten percent of those e-mails are forwarded at least once.

(4) At the Scarborough Hospital in Tobago, only 5% of newborn babies are born on their due date. Find the probability that more than 6% of the next 150 babies delivered are born on their due date.

(5) A certain gated community allows residents a choice of either a dog or a cat as a pet. If the number of families who have a dog is equal to the number of families who have a cat, determine the probability that from a random sample of 51 residents, the proportion who chose dogs as pets would be greater than one-third but less than three-fifths.

(6) Ninety-five percent of households in the Cuban capital Havana use fluorescent light bulbs. The Ministry of Energy decides to conduct a random inspection of Havana households. If a sample of 40 households is selected for inspection, what is the likelihood that less than eighty percent of these 40 households would be using fluorescent light bulbs?

(7) Two-thirds of the viewers of a popular TV talk show are women. If 35 of the viewers of this talk show are randomly selected for a marketing promotion, how probable is it that less than half of them are women?

(8) During the outbreak of bird flu, it was estimated that 45% of all birds on farms were infected. Health inspectors decide to test randomly selected birds from across the country. What is the probability that more than half of a random sample of 200 birds would be infected by the bird flu?

(9) 60% of the credit card customers of a certain bank in Lagos, Nigeria use Mastercard. If 75 of the bank's customers are randomly selected, what is the probability that between 50% and 55% of the sample **DO NOT** use Mastercard.

(10) In Trinidad and Tobago, there are two mobile phone providers – Digicel and bmobile. 55% percent of mobile phone customers use bmobile. Assume that each mobile phone user subscribes only to one provider. Find then the probability that between 40% and 45% of a random sample of 120 mobile phone users in Trinidad and Tobago have Digicel phones.

6
Confidence Intervals

More often than not it is extremely inconvenient or practically impossible to measure a given characteristic for *every* member of a population. A sample is therefore taken, and a **sample statistic** is used as an estimate of the corresponding **population parameter**. For example, the mean of a sample is used to estimate the mean of the population, or the proportion of a sample that possesses a particular characteristic is used to estimate the proportion of the population that possesses that particular characteristic. This approach is called the **point estimate** approach, where we simply take the value of the sample statistic as an estimate of the population parameter. For example, we may want to calculate the mean number of calories consumed daily by the population of a particular town. Instead of measuring the caloric intake of every single resident of the town, we take a sample of say, 100 residents and find the mean caloric intake for that particular sample. We then use the mean caloric intake of our particular sample as an estimate of the mean caloric intake of the entire population of the town in question. As another example, we may want to estimate the proportion of families in this same town that have more than three children. Instead of taking a census of every family in the town, we take a sample of say, 150 families, and then from this sample, we measure the proportion of families that have more than three children. We then use this figure as an estimate of the proportion of families in the entire town with more than three children. In general, the following sample statistics are used as point estimates for the corresponding population parameters:

Sample Statistic	*Population Parameter*
Sample Mean	Population Mean
Sample Proportion	Population Proportion
Sample Variance	Population Variance
Sample Standard Deviation	Population Standard Deviation

The problem with the point estimate approach is that any sample we take is but one of literally *millions* of possible samples of the same size that can be drawn from a given population. We cannot then take the results of our singular sample to be a strict and accurate estimate of the population parameter. The point estimate approach on its own is therefore insufficient, because the value of the sample statistic we obtain would depend on the particular sample taken. So for example, we may measure the mean of a sample of size 50, and get a result of 45. The sample mean is 45. Using the point estimate approach, we would simply say that the population mean is 45, based on our sample result. But another sample of size 50 may have a mean of 41, and another sample of the same size may have a mean of 47, and so on. This limitation of the point estimate approach is compensated for by using the **interval estimate** approach.

The interval estimate is simply an interval which contains an innumerable multitude of point estimates, and the interval estimate approach utilizes the interval to give us some 'breathing room' with respect to the population parameter whose value we are using the sample statistic to estimate. With an interval estimate, we give ourselves a much greater chance of including the actual value of the population mean, as opposed to when we use a point estimate. So we use the sample mean to construct an interval, and we can then say for example, that 'the population mean lies between 45 and 55'. If 95% of the available point estimates lie between 45 and 55, then the interval between (and including) 45 and 55 is our '95% Confidence Interval' for the population mean. Similarly, an interval that contains 99% of the possible point estimates based on a sample of a particular size is referred to as a '99% Confidence Interval'. Confidence Intervals can take any size, but in practice, the most useful Confidence Intervals are those that are at least 80%.

We may take a sample of size 80 and measure the existence of a particular characteristic. We come up with 20 members of the sample having that particular characteristic. This means that the proportion of the sample that possesses the characteristic in question is $\frac{20}{80}$ (0.25). Using the point estimate approach, we would simply say that 25 percent of the population possesses this characteristic. In other words, we would be saying that the proportion of the population possessing that characteristic is 0.25. But another sample of size 80 may give a proportion of 0.22, and yet another sample of the same size may give a proportion of 0.28. We therefore find an interval within which we may say that 95% or 99% of the available point estimates for the population proportion would lie. These would be examples of a 95% and a 99% confidence interval for the population proportion.

Although the theoretical interpretation of the confidence interval takes into account the point estimates taken from *all* samples of a specified size, practically speaking, we only need one sample to construct a working confidence interval. In the previous chapter, we learnt how to construct sampling distributions for the sample mean and the sample proportion. It is the sampling distributions that we use, *at all times*, to construct confidence intervals. We shall here consider only confidence intervals involving the sample mean and the sample proportion, and we shall deal only with large samples.

*Confidence Intervals are **ALWAYS** constructed using the Sampling Distributions!!*

ESTIMATING THE POPULATION MEAN
Confidence Interval for the Population Mean
In chapter 5, we learnt that the sampling distributions are constructed with due consideration for the availability of parameters. Once we have the population parameters available, we use them, and when we don't, we use the sample statistics.

We calculate confidence intervals for the population mean because we don't know its precise value, which means that the population mean will never be available to us when constructing confidence intervals for the population mean. We will always be using the sample mean as our starting point. The population variance may or may not be available.

Samples drawn from Normal Populations (*known population variance*)

Consider the class of 100 university students from chapter 5. Recall that these students buy fast food every day at Top Lunch Café. We wish to measure the mean expenditure on fast food for this class of 100. Let us suppose that the standard deviation of the expenditure on fast food for this entire population of 100 students is known to be 5 dollars. We also know that the expenditure on fast food for the 100 students is normally distributed. We therefore have a random variable 'the mean expenditure on fast food for a class of 100 university students', which is normally distributed with a mean μ (which we are trying to estimate), and a known standard deviation of 5.

We wish to estimate this mean for the population of 100 students, but having neither the time, the resources, nor the inclination to conduct a full-scale census, we take a sample of size 40. From this sample we get a mean expenditure of $25 for the fast food. Do we then take this as an estimate for the entire population? Yes we can, but in doing so, we must remain cognizant of the fact that our sample is just ONE of a very large number of possible samples of size 40 from a population of 100. As mentioned in chapter 5, the number of samples of size 40 that can be drawn from a population of 100 is equal to 13,746,234,150,000,000,000,000,000,000, or 13.75 octillion! So our lone sample of size 40 is but just one of a total of 13.75 octillion possible samples of size 40 from a population of 100! Imagine then how many possible samples we can have if the population was, say, 1,000! This is the reason we have confidence intervals, because the sample mean that we obtain is simply a function of which particular sample we use.

It is here that we start to consider the Distribution of the Sample Mean. According to the theory we mentioned in the last chapter, if we were to take all 13.75 octillion samples and measure their means, we would find that those means are distributed normally with a mean equal to the population mean μ, and a variance equal to the population variance divided by the sample size $\dfrac{\sigma^2}{n}$ (standard deviation $\dfrac{\sigma}{\sqrt{n}}$). We can therefore use the sample mean of 25 as a point estimate for the population mean in formulating the distribution of the sample mean. We know the population standard deviation is 5, so therefore the variance of the distribution of the sample mean would be $\dfrac{5^2}{40}$, and the standard deviation is $\dfrac{5}{\sqrt{40}}$.

$$\therefore \ \overline{X} \sim N\left(25, \frac{5^2}{40}\right)$$

$$\Rightarrow \overline{X} \sim N(25, 0.625)$$

The mean of the distribution of the sample mean for samples of size 40, based on our sample of 40 is 25, and the variance is 0.625. Hence the standard deviation is $\sqrt{0.625} = 0.7906$. This is the distribution that we would use to calculate the confidence interval for the population mean μ. Let us now calculate a 95% confidence interval.

We represent the distribution of the sample mean for samples of size 40 on the Normal Distribution curve, so that when we seek to construct a 95% confidence interval, what we are actually looking for is an interval between which 95% of the means of *all* the samples of size 40 would lie. Represented graphically, what we seek is:

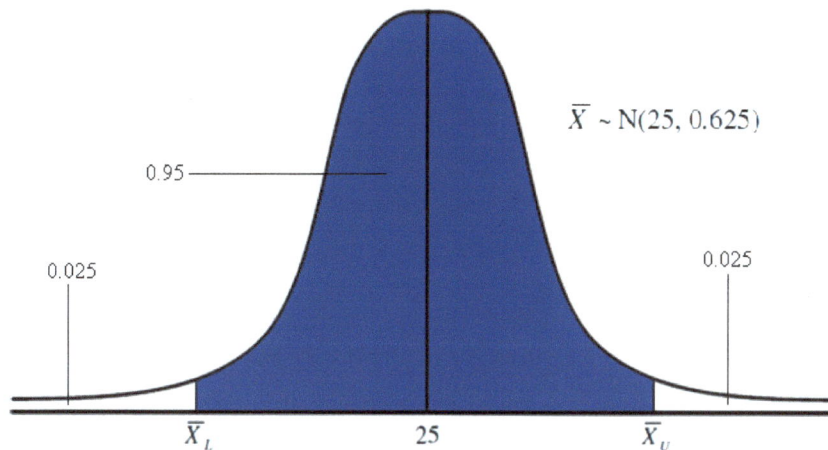

$$\overline{X} \sim N(25, 0.625)$$

We want two values $\overline{X}_L$ and $\overline{X}_U$ such that the probability that a sample mean from a sample of size 40 would lie between these two values is 0.95. What this means is that the probability that the sample mean from a sample of size 40 would lie outside this range is 0.05. From the graph, we can see that the probability that the sample mean would be less than the lower limit of the interval $\overline{X}_L$ is 0.025 (half of 0.05), and the probability that it would be greater than the upper limit of the interval $\overline{X}_U$ is also 0.025. So how do we proceed?

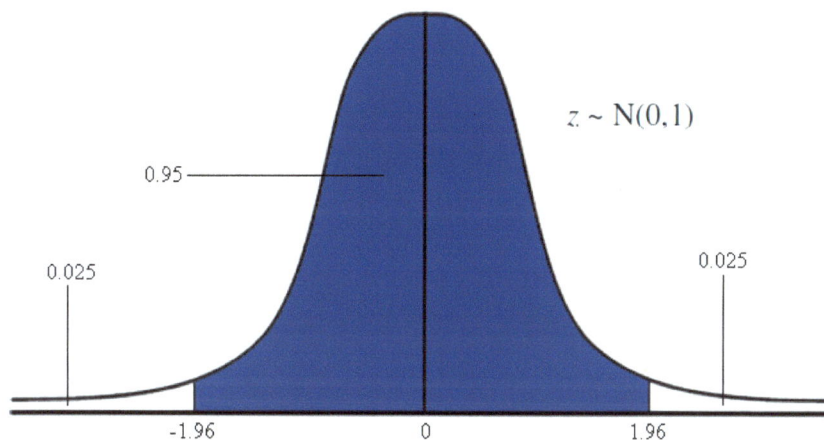

$$z \sim N(0,1)$$

We use the z-table to find the z-values that would correspond to probabilities of 0.025 in the left tail and the right tails of the standard normal curve. The required z-values are -1.96 for the left tail, and 1.96 for the right tail. We now proceed to de-standardize in order to find the values of the sample mean that correspond to the z-values of -1.96 and 1.96. These would give us the lower and upper limits of the interval within which 95% of the means of random samples of size 40 would lie. So:

$$-1.96 = \frac{\overline{X}_L - 25}{0.7906} \qquad\qquad 1.96 = \frac{\overline{X}_U - 25}{0.7906}$$

Cross-multiplying:

$$(-1.96)(0.7906) = \overline{X}_L - 25 \qquad\qquad (1.96)(0.7906) = \overline{X}_U - 25$$
$$-1.550 = \overline{X}_L - 25 \qquad\qquad 1.550 = \overline{X}_U - 25$$
$$25 - 1.550 = \overline{X}_L \qquad\qquad 25 + 1.550 = \overline{X}_U$$
$$\therefore \quad \overline{X}_L = 23.45 \qquad\qquad \therefore \quad \overline{X}_U = 26.55$$

Therefore, based on this particular sample, the 95% confidence interval for the population mean would be *23.45 ≤ μ ≤ 26.55*. What this means is that if somehow we were able to measure the means of all 13.75 octillion possible samples of size 40 from our population of 100, 95% of those sample means would lie between 23.45 and 26.55. In other words, 95% of all the possible point estimators for the population mean based on samples of size 40 would lie between 23.45 and 26.55. We can therefore say that based on all the possible point estimates from samples of size 40, the probability that the population mean lies within the interval *23.45 ≤ μ ≤ 26.55* is 0.95. This result is alternatively interpreted to mean that based on this sample of 40, we can say with 95% confidence that the population mean will lie between 23.45 and 26.55.

Keep in mind of course, that a different sample of size 40 may yield a different mean, and hence a different confidence interval. Bear in mind also that samples of a different size would yield different distributions of the sample mean, and hence different confidence intervals. In general, larger samples tend to give smaller, more precise intervals for any particular confidence level.

Samples drawn from Normal Populations *(unknown population variance)*

Consider again the class of 100 university students from the last example. Suppose, as usually happens in real life, that we have a situation where, apart from not knowing the population mean, we also do not know the population variance. In addition to the population mean, the population standard deviation is also unknown. How do we then proceed?

Suppose we take a sample of size 40 as before. This time we get a sample mean of 26, and a sample standard deviation of 4 (variance = 4^2 = 16). As in the previous case, we use the sample mean of 26 as a point estimate for the population mean. In this instance, however, the population standard deviation is unknown. We therefore use the sample standard deviation as a point estimate of the population standard deviation, and then apply the Central Limit Theorem as before.

The distribution of the sample mean in this case would have a mean of 26, and a variance of $\dfrac{4^2}{40}$. Therefore, $\overline{X} \sim N(26, \dfrac{4^2}{40})$, $\Rightarrow \overline{X} \sim N(26, 0.4)$. The standard deviation of this distribution is $\sqrt{0.4} = 0.632$. This is the distribution that we would use to calculate the confidence interval for the population mean 'μ'. Let us now construct a 96% confidence interval.

We represent the distribution of the sample mean for samples of sizes of size 40 on the Normal Distribution curve, and then we seek to construct a 96% confidence interval, so what we are actually looking for is an interval between which 96% of the means for samples of size 40 would lie. Represented graphically, what we seek is:

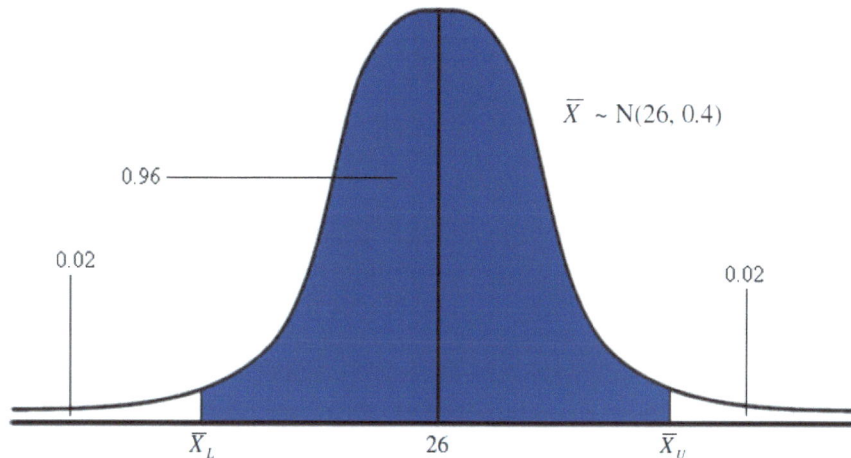

We want two values $\overline{X}_L$ and $\overline{X}_U$ such that the probability that a sample mean from a sample of size 40 would lie between these two values is 0.96. What this means is that the probability that the sample mean from a sample of size 40 would lie outside this range is 0.04. From the graph, we can see that the probability that the sample mean would be less than the lower limit of the interval $\overline{X}_L$ is 0.02 (half of 0.04), and the probability that it would be greater than the upper limit of the interval $\overline{X}_U$ is also 0.02. We then use the z-table to find the z-values that would correspond to probabilities of 0.02 in the left tail and the right tail of the standard normal curve.

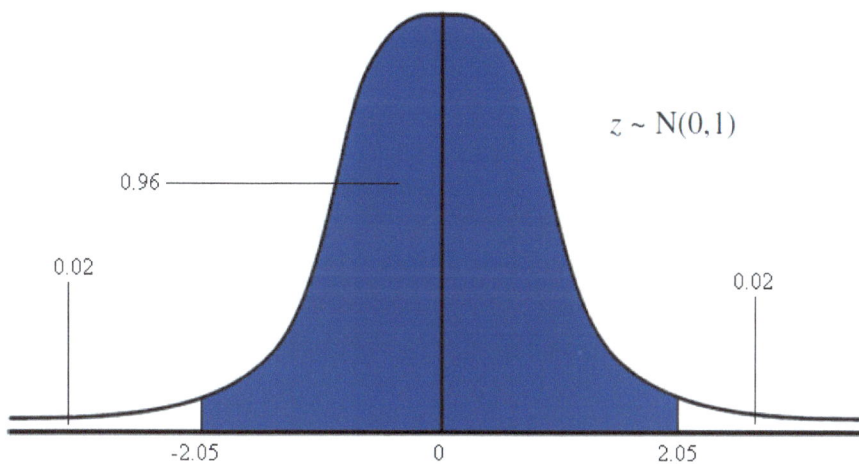

The values that correspond to these areas are -2.05 for the left tail and 2.05 for the right tail. We now need to de-standardize in order to find the values of the $\overline{X}$ variable that correspond to the z-values of -2.05 and 2.05. These would give us the lower and upper limits of the interval within which 96% of the means of random samples of size 40 would lie.

So:

$$2.05 = \frac{\overline{X}_L - 26}{0.6325} \qquad\qquad 2.05 = \frac{\overline{X}_U - 26}{0.6325}$$

$$(-2.05)(0.6325) = \overline{X}_L - 26 \qquad\qquad (2.05)(0.6325) = \overline{X}_U - 26$$

$$-1.297 = \overline{X}_L - 26 \qquad\qquad 1.297 = \overline{X}_U - 26$$

$$26 - 1.297 = \overline{X}_L \qquad\qquad 26 + 1.297 = \overline{X}_U$$

$$\therefore \quad \overline{X}_L = 24.703 \qquad\qquad \therefore \quad \overline{X}_U = 27.297$$

Therefore, based on this particular sample the 96% confidence interval for the population mean would be *24.703 ≤ μ ≤ 27.297*. What this means is that if somehow we were able to measure the means of all 13.75 octillion possible samples of size 40 from our population of 100, 96% of those sample means would lie between 24.703 and 27.297. This means that 96% of all the possible point estimates for the population mean based on samples of size 40 would lie between 24.703 and 27.297. We can therefore say that, based on all the possible point estimates from samples of size 40, the probability that the population mean lies within the interval *24.703 ≤ μ ≤ 27.297* is 0.96. This result is alternatively interpreted to mean that based on this sample of 40, we can say with 96% confidence that the population mean will lie between 24.703 and 27.297

Large Samples drawn from Non-Normal or Unknown Populations

We have just covered the cases where the sample was drawn from populations that are known to be distributed normally. In the case where the population distribution is known to be non-normal, or is simply unknown, we can invoke the Central Limit Theorem, as long as the sample is large (sample size greater than 30).

So for example, we may have a situation where the mean expenditure on lunch for the entire population of 100 university students in a class is unknown. The standard deviation of the costs of these lunches is also unknown. The expenditure on lunches for the entire class may be known to not be a normal distribution, or the distribution of the expenditure on lunches may be simply unknown. We wish to develop a confidence interval for the mean expenditure on lunches. Our only option is to use a sample. Let us a use a sample of size 81, and suppose the sample mean is 28. Suppose also that the sample standard deviation is 7.

Because the sample is large (*n > 30*), we are free to use the Central Limit Theorem to develop a distribution of the sample mean that would be normal. Our sample mean is 28, so we use this value as a point estimate of the value of the population mean. We use the value of the sample standard deviation of 7 as a point estimate of the value of the population standard deviation. Having done this, we then say that the mean of the distribution of the sample mean is 28, with a variance of $\frac{7}{81}$ (standard deviation $\frac{7}{\sqrt{81}}$), and we proceed EXACTLY as we did in the cases where the sample was drawn from a Normal Population. As an exercise, construct a 90% Confidence Interval in this present scenario.
Answer: *27.01 ≤ μ ≤ 28.99*

Exercise 6.1

Construct the required confidence intervals:

(1) If $Y \sim N(\mu, 11.1^2)$, and a sample of size 44 drawn from this population has a mean of 56.3, construct the following confidence intervals:

(a) 91%　　　　(b) 89%　　　　(c) 94%　　　　(d) 99%　　　　(e) 92%

(2) If $W \sim N(\mu, \sigma^2)$, and a sample of size 112 drawn from this population has a mean of 33.54, and a standard deviation of 9.22, construct the following confidence intervals:

(a) 99%　　　　(b) 98%　　　　(c) 96%　　　　(d) 85%　　　　(e) 83%

(3) The average height of a sample of 57 trees from a rainforest is 4,563 cm. The standard deviation of the heights of this sample of trees is 225 cm. Construct a 95% confidence interval for the mean height of trees in the entire rainforest.

(4) The average daily temperature during the month of March in a certain Caribbean island is 31.6 degrees Celsius, with a standard deviation of 3.4 degrees Celsius. Use the data for March to construct a 97% confidence interval for the average daily temperature of the island.

(5) A sample of 46 families in a particular neighborhood of a large city revealed that the average time spent looking at television each day was 2.5 hours. The standard deviation from this sample was calculated to be 0.5 hours. Use this data to construct a 90% confidence interval for the amount of time that families in the entire city spend looking at television.

(6) A glossy fashion magazine has measured the times spent by a sample of 36 female university students to get dressed. The magazine found that the average time to get dressed among these students was 45 minutes, with a standard deviation of 9 minutes. Use the data from this sample to construct a 96% confidence interval for the average time taken by any female student of that university to get dressed.

(7) A particular pastor has just concluded his 52nd consecutive weekend sermon at the neighborhood church. A particular church-goer, an avid student of statistics, has measured the time taken for each one of these 52 sermons, and wishes to use this data as a basis for estimating the average length of the pastor's sermons. His calculations reveal that the mean length of these 52 sermons was 77.77 minutes, with a standard deviation of 12.32 minutes. If he decided to construct a 99% confidence interval for the length of the pastor's sermons, what would that interval be?

(8) The weights of a sample of 100 newborns at a regional hospital are normally distributed with a mean of 7.8 ounces and a standard deviation of 1.23 ounces. Using this sample, construct a 98% confidence interval for the mean weight of all the newborns at this hospital.

(9) The quality control department at a local distillery wishes to know how much beer is deposited into the empty bottles at the end of the production process. A random sample of 225 bottles is selected, and the volume of beer in each bottle measured. The mean volume of beer in each bottle is found to be 324.85ml, with a standard deviation of 1.33ml. Develop a 99% confidence interval for the mean volume of beer deposited in all bottles in the distillery.

(10) The average wingspan of a sample of 25 adult birds of a particular species is 1.35m. The standard deviation of this sample is found to be 0.1m. If the wingspans of the birds of this species is normally distributed, what is the 94% confidence interval for the mean wingspan of birds of this species?

One-Sided Confidence Intervals for the Population Mean

Unless specifically instructed otherwise, the use of the term 'confidence interval' usually refers to a two-sided confidence interval, where we have a lower boundary and an upper boundary. However, one-sided confidence intervals with only a lower limit or an upper limit do in fact exist.

Upper Confidence Intervals

The confidence intervals covered thus far are examples of two-sided confidence intervals. Suppose instead that we want to find a 95% confidence interval for the **minimum value** of the mean expenditure on fast food for the population of 100 university students. What this means is that the 95% area under consideration would be confined to the upper portion of the normal curve instead of being spread equally about the mean as is the case for two-sided confidence intervals. This is the reason why a confidence interval for the minimum value is called an upper confidence interval.

As in the last section, we use samples of size 40. In this instance we want only a minimum value above which 95% of the means for samples of size 40 would lie. Therefore the probability that the mean is less than this minimum value would be 0.05 $(1 - 0.95)$. Graphically, this situation would appear thus:

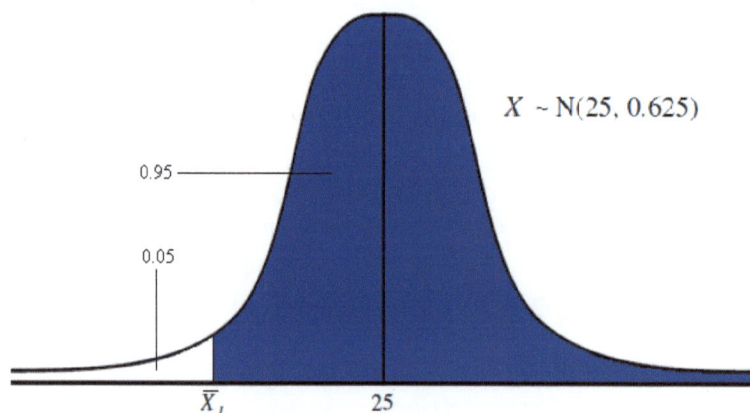

We now go to the z-table and look for the z-value that gives an area of 0.05 in the left tail of the standard normal curve:

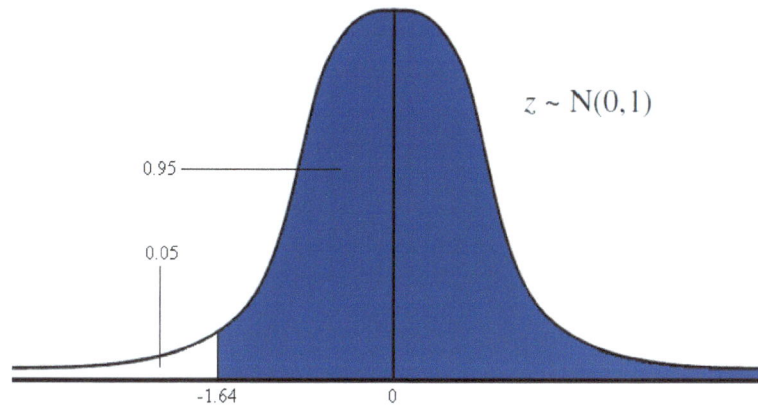

From the z-table, this value is -1.64, so we need to de-standardize this z-value to find the value of $\overline{X}_L$ on the distribution of the sample mean that would correspond to the lower boundary of our confidence interval. So:

$$-1.64 = \frac{\overline{X}_L - 25}{0.7906}$$

$$(-1.64)(0.7906) = \overline{X}_L - 25$$

$$-1.305 = \overline{X}_L - 25$$

$$25 - 1.305 = \overline{X}_L$$

$$\therefore \quad \overline{X}_L = 23.695$$

So, we can say that the upper 95% confidence interval for the value of the mean cost of food for the entire population is: $\mu \geq 23.695$. Based on this particular sample, we can say that if the means of all the possible samples of size 40 were tabulated, 95% of those means would be greater than 23.695. This means that 95% of all the possible point estimates of the population mean based on samples of size 40 would be greater than 23.695. We can therefore say that, based on all the possible point estimates from samples of size 40, the probability that the population mean lies within the interval $\mu \geq 23.695$ is 0.95. This result is alternatively interpreted to mean that based on this sample of 40, we can say with 95% confidence that the population mean would be greater than 23.695.

Lower Confidence Intervals

Let us now reverse the situation. Suppose instead that we wish to find a lower 95% confidence interval for the **maximum value** of the mean expenditure on fast food for the population of 100 university students. What this means is that the 95% area under consideration would be confined to the lower portion of the normal curve instead of being spread equally about the mean as is the case for two-sided confidence intervals. This is the reason why a confidence interval for the maximum value is called a lower confidence interval.

Again we use a sample of size 40. What we are in fact seeking is a value below which 95% of the possible point estimators would lie. And we get this value based on our sample of size 40. So what we have graphically is:

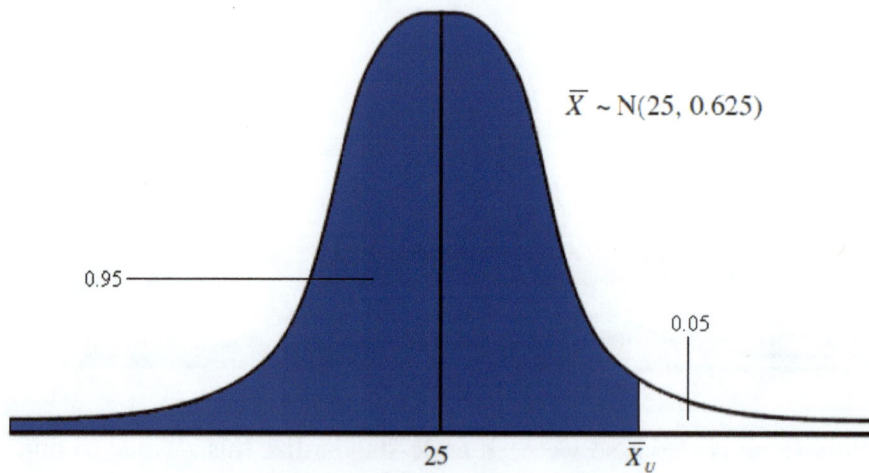

$\overline{X} \sim N(25, 0.625)$

0.95

0.05

25     $\overline{X}_U$

We now go to the z-table to find the standardized z-value that gives an area of 0.05 in the right tail of the z-curve. This value is 1.64:

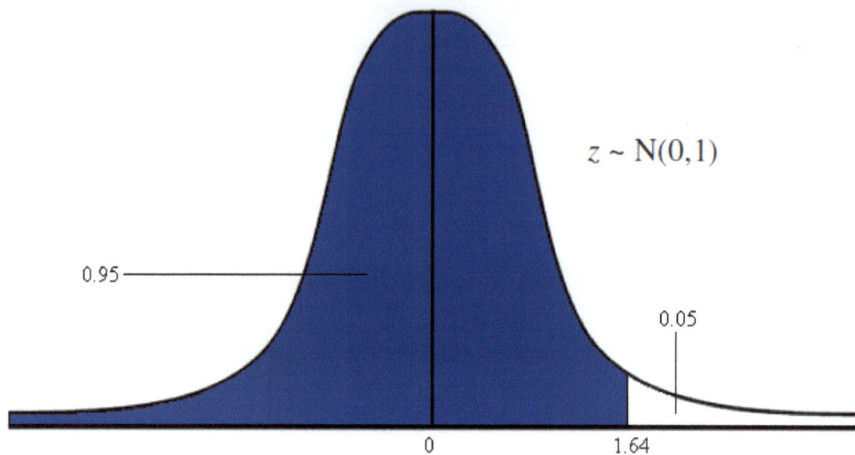

$z \sim N(0,1)$

0.95

0.05

0   1.64

We need to de-standardize this z-value to find the value of $\overline{X}_U$ on the distribution of the sample mean that would correspond to the maximum value of our confidence interval. So,

$$1.64 = \frac{\overline{X}_U - 25}{0.7906}$$

$$(1.64)(0.7906) = \overline{X}_L - 25$$

$$1.305 = \overline{X}_U - 25$$

$$25 + 1.305 = \overline{X}_U$$

$$\therefore \quad \overline{X}_U = 25.305$$

So, we can say that the lower 95% confidence interval for the value of the mean cost of food for the entire population is: $\mu \leq 25.305$. Based on this particular sample, we can say that if the means of all the possible samples of size 40 were tabulated, 95% of those means would be less than 25.305. This means that 95% of all the possible point estimators for the population mean based on samples of size 40 would be less than 25.305. We can therefore say that, based on all the possible point estimators from samples of size 40, the probability that the population mean lies within the interval $\mu \leq 25.305$ is 0.95. This result is alternatively interpreted to mean that based on this sample of 40, we can say with 95% confidence that the population mean would be less than 25.305.

Remember, when we know the population variance, we use it straightaway in forming that distribution of the sample mean that we would use to calculate the confidence interval. When the population variance is unknown, we use the sample variance as a point estimator of the population variance. From this we construct the distribution of the sample mean, which is used to construct the required confidence interval.

Now repeat these same procedures for the same 95% one-sided confidence intervals for the case where we have a sample of size 40 drawn from a non-normal or unknown population where the sample mean is 26 and the sample standard deviation is 4.
(Lower Confidence Interval: $\mu \leq 26.8095$) (Upper Confidence Interval: $\mu \geq 25.1905$)

Exercise 6.2
For each of the following situations, construct the requested one-sided confidence interval:

(1) If $\bar{B} \sim N(35, 5^2)$, construct the following upper confidence intervals for the population mean of B from a sample of size 121:
(a) 90%   (b) 93%   (c) 89%   (d) 99%   (e) 95%   (f) 91%

(2) If $\bar{T} \sim N(67, 6.54^2)$, construct the following lower confidence intervals for the   population mean of T from a sample of size 16, given that the variable 'T' is normally distributed:
(a) 86%   (b) 89%   (c) 92%   (d) 94%   (e) 96%   (f) 98%

(3) The average height of a sample of 57 trees from a rainforest is 4,563 cm. The standard deviation of the mean height of this sample of trees is 225 cm. Construct a 95% upper confidence interval for the minimum value of the mean height of trees in the   entire rainforest.

(4) The average daily temperature during the month of March in a certain Caribbean island is 31.6 degrees Celsius, with a standard deviation of 3.4 degrees Celsius. Use the data for March to construct a 97% lower confidence interval for the maximum value of the average temperature of the island.

(5) A sample of 46 families in a particular neighborhood of a large city found that the average time spent looking at television each day was 2.5 hours. The standard deviation from this sample was calculated to be 0.5 hours. Use this data to construct a 93% lower confidence interval for the maximum value of the average time that families in the entire city spend looking at television.

(6) A glossy fashion magazine has measured the times spent by a sample of 36 female university students to get dressed. The magazine found that the average time to get dressed among these students was 45 minutes, with a standard deviation of 9 minutes. Use the data from this sample to construct a 89% upper confidence interval for the minimum value of the average time taken by any female student of that university to families in the entire city spend.

(7) A particular pastor has just concluded his 52nd consecutive weekend sermon at the neighborhood church. Yolande, a member of this church, an avid student of statistics, has measured the time taken for each one of these 52 sermons, and wishes to use this data as a basis for estimating the average length of the pastor's sermons. Her calculations reveal that the mean length of these 52 sermons was 77.77 minutes, and that the standard deviation of the lengths of these sermons was 12.32 minutes. If she decided to construct a 94% lower confidence interval for the maximum value of the average length of the pastor's sermons, what would that interval be?

(8) The weights of a sample of 100 newborns at a regional hospital are normally distributed with a mean of 7.8 ounces and a standard deviation of 1.23 ounces. Using this sample, construct an 88% upper confidence interval for the minimum value of the mean weight of all the newborns at this hospital.

(9) The quality control department at a local distillery wishes to know how much beer is deposited into the empty bottles at the end of the production process. A random sample of 225 bottles is selected, and the volume of beer in each bottle measured. The mean volume of beer in each bottle is found to be 324.85ml, with a standard deviation of 1.33 ml. Develop a 93% lower confidence interval for the maximum value of the mean volume of beer deposited in all bottles in the distillery.

(10) The average wingspan of a sample of 25 adult birds of a particular species is 1.35m. The standard deviation of this sample is found to be 0.1m. If the wingspans of the birds of this species are normally distributed, what is the 92% upper confidence interval for the minimum value of the average wingspan of birds of this species?

Confidence Interval for the Difference Between Two Population Means

Very often we need to use two samples to construct a confidence interval for the difference between the means of the populations from which the samples were drawn. We do so by using the sample means and their corresponding sample variances/standard deviations. We will restrict our treatment of confidence intervals for the difference between two population means to those cases where both samples are large *(n > 30)*. The underlying assumption is that the samples are independent of each other.

For example, consider the following scenario:

Example 6.1

A sample of 55 European films was found to have a mean length of 105 minutes with a standard deviation of 7 minutes. A sample of 62 Hollywood films was found to have a mean length of 90 minutes with a standard deviation of 5 minutes. Determine a 95% confidence interval for the difference between the length of a European film and the length of a Hollywood film.

Solution

For both samples, the distribution of the population is unknown, but since the sample size is greater than 30 in both cases, we have statistically large samples. The distributions of these sample means would therefore follow the normal distribution, with means equal to 105 and 90 minutes respectively for European and Hollywood films.

In both cases, the population variance is unknown, so we use the next available option, which is the variance of the respective samples. In each case, the sample variance is used as a point estimate of the population variance. We then obtain the variance of each sample distribution by dividing the population variance by the sample size. So the respective variances would be $\dfrac{7^2}{55}$ for European films, and $\dfrac{5^2}{62}$ for Hollywood films.

As always, we first define the random variables:

Let $\overline{E}$ be the randon variable 'mean length of a sample of 55 European films'.

$$\overline{E} \sim N(105, \frac{7^2}{55})$$

Let $\overline{H}$ be the random variable 'mean length of a sample of 62 Hollywood films'.

$$\overline{H} \sim N(90, \frac{5^2}{62})$$

Before we do anything else, we must first construct a new normal distribution for the difference between the means of the samples of European and Hollywood films.

Let $\overline{E} - \overline{H}$ be the random variable *'difference between the average lengths of European and Hollywood films'*. We recall from chapter 4 that when constructing a normal distribution for the difference between two normal variables, we take the difference between the means, and we add the variances. So, in this case:

$$E(\overline{E} - \overline{H}) = E(\overline{E}) - E(\overline{H})$$
$$= 105 - 90$$
$$= 15$$
$$Var(\overline{E} - \overline{H}) = Var(\overline{E}) + Var(\overline{H})$$
$$= \frac{7^2}{55} + \frac{5^2}{62}$$
$$= 0.8909 + 0.4032$$
$$= 1.2941$$

$\therefore \overline{E} - \overline{H} \sim N(15, 1.2941)$

We now proceed with the construction of the confidence interval as we did previously.

We let $D = \overline{E} - \overline{H}$

Therefore, $D \sim N(15, 1.2941)$

We can represent the distribution for D on a normal curve, so that when we seek to construct a 95% confidence interval, what we are actually looking for is an interval between which 95% of the differences between the means of European and Hollywood films would lie. Represented graphically, what we seek is:

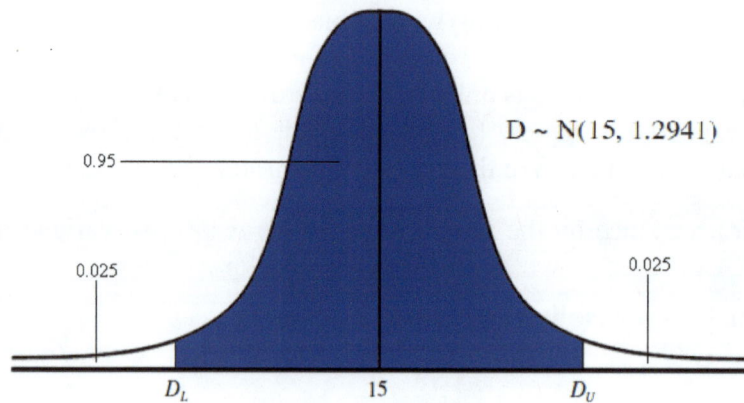

$D \sim N(15, 1.2941)$

D_L is the lower limit of this interval, while D_U is the upper limit. The next step is to go to the standard normal curve and locate the z-values that would correspond to the values D_L and D_U. These would be the z-values that would give areas of 0.025 in the extreme right and left tails of the standard normal curve. These values are -1.96 and 1.96. On the curve, we have:

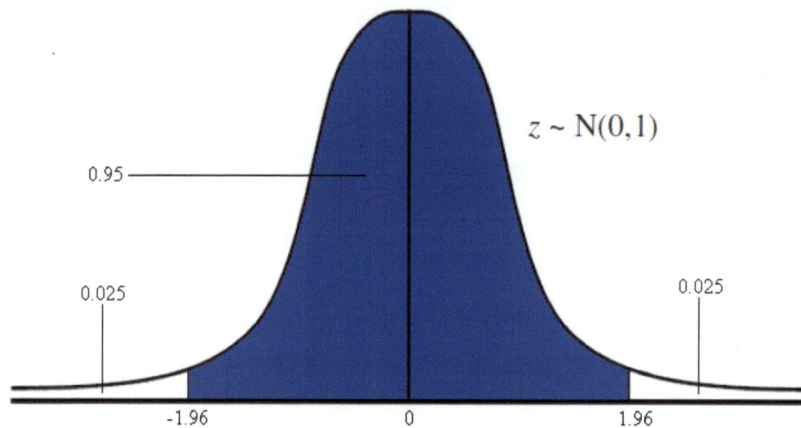

$z \sim N(0,1)$

Therefore, we now de-standardize to find the values D_L and D_U that correspond to -1.96 and 1.96 respectively.

$$-1.96 = \frac{D_L - 15}{\sqrt{1.2941}}$$

$$= \frac{D_L - 15}{1.1376}$$

$$(-1.96)(1.1376) = D_L - 15$$
$$-2.230 = D_L - 15$$
$$15 - 2.230 = D_L$$
$$\therefore \quad D_L = 12.77$$

$$1.96 = \frac{D_U - 15}{\sqrt{1.2941}}$$

$$= \frac{D_U - 15}{1.1376}$$

$$(1.96)(1.1376) = D_U - 15$$
$$2.230 = D_U - 15$$
$$15 + 2.230 = D_U$$
$$\therefore \quad D_U = 17.23$$

Therefore the 95% confidence interval for the difference between the mean lengths of European and Hollywood films in minutes is (12.77, 17.23)

What this result means is that if all possible combinations of a random sample of 55 European films and a random sample of 62 Hollywood films are considered and the means compared, 95% of the time, the mean for European films will be greater than the mean for Hollywood films by a value that lies between 12.77 and 17.23 minutes. In other words, the probability that the mean length of 55 random European films will be greater than the mean length of 62 random Hollywood films by between 12.77 and 17.23 minutes is 0.95. Alternatively, the result is expressed by saying that we are 95% percent certain that the difference between a European film and a Hollywood film lies between 12.77 and 17.23 minutes.

Exercise 6.3

(1) The maximum night-time temperature in the borough of Arima over the course of 23 days during the dry season in Trinidad and Tobago was measured and found to have a mean of 27 degrees Celsius and a variance of 2.25 degrees Celsius. Similarly the maximum night-time temperature in the borough of Point Fortin over the same 23 days was measured and found to have a mean of 26 degrees Celsius and a standard deviation of 1.23 degrees Celsius. It is known that the temperature measurements for the boroughs of Arima and Point Fortin are normally distributed. Develop a 97% confidence interval for the difference between the average maximum night-time temperatures in Arima and Point Fortin.

(2) 100 newborn babies at the San Fernando General Hospital have an average weight of 7.23 pounds. The standard deviation of these weights is 1.34 pounds. 86 newborns at the Port-of-Spain General Hospital have an average weight of 7.06 pounds, the standard deviation of their weight measurements being 0.98 pounds. Use these measurements to construct a 95% confidence interval for the difference between the mean weights of newborns at the Port-of-Spain and San Fernando General Hospitals.

(3) The average earthquake intensity over the course of 45 days in the Vanuatu Region was found to be 4.93 on the Richter Scale. The variance of these measurements was found to be 1.98. The average earthquake intensity in the Puerto Rico Region over a period of 57 days was found to be 3.81 on the Richter Scale with a variance of 1.65. Use this information to construct a 93% confidence interval for the difference between the average earthquake intensities in the Vanuatu Region and the Puerto Rico Region.

(4) A sample of 99 boys at the Aga Khan High School in Uganda have an average height of 179.4 cm, with a standard deviation of 7.66cm. A sample of 77 boys from the Jabulani Technical High School in South Africa have an average height of 165.7 cm, with a standard deviation of 5.88cm. Construct a 96% confidence interval for the difference between the average heights of boys at the Aga Khan High School and boys at the Jabulani Technical High School.

(5) A sample of 123 girls at the Belair High School in Jamaica have a mean weight of 135 pounds with a variance of 27.3 pounds. A similar sample of 101 girls at Queen's College in Guyana have a mean weight of 138 pounds with a standard deviation of 4.63 pounds. Develop a 99% confidence interval for the difference between the mean weights of girls at Belair High School and girls at Queen's College

(6) The records of 90 randomly chosen patients at a particular hospital are examined. The mean systolic pressure of these patients is 154.3 mmHg, with a standard deviation of 9.09 mmHg. The records of another 90 patients are randomly chosen and the diastolic pressures from these records are analysed. The mean diastolic pressure is 96.7 mmHg and the variance of these pressures is 50.24 mmHg. Construct a 97% confidence interval for the difference between the mean systolic and the mean diastolic pressures for all the patients in the hospital.

(7) The mean time for a sample of 60 randomly selected flights from Port-of-Spain to London is 7.63 hours, with a standard deviation of 0.2 hours. The mean time for a sample of 75 randomly selected return flights from London to Port-of-Spain is 8.07 hours with a standard deviation of 0.32 hours. Develop a 92% confidence interval for the difference between the average flight time from Port-of-Spain to London and the average flight time from London to Port-of-Spain.

(8) The average weight of a sample of 45 persons entering a weight-loss clinic is 302.7 pounds, with a variance of 225.87 pounds. The average weight of another sample of 67 persons leaving the clinic upon completion of their program is 207 pounds with a variance of 82.3 pounds. Use this data to construct a 98% confidence interval for the difference between the average weight of a patient before the weight-loss program, and the average weight of a patient after the weight-loss program.

(9) The average reaction time of a sample of 32 elite 100 meter sprinters before the introduction of false-start technology in 2003 was 0.145 seconds with a standard deviation of 0.022 seconds. The mean reaction time for another sample of 32 elite 100 meter sprinters after the introduction of false-start technology had a mean of 0.156 seconds with a standard deviation of 0.026 seconds. Use this data to develop a 98% confidence interval for the difference in reaction times of elite 100 meter sprinters before the introduction of false-start technology and the reaction times after the introduction of false-start technology.

(10) The mean length of a random sample of 567 phone calls from the local mobile provider was found to be 2.10 minutes with a standard deviation of 0.56 minutes. In order to induce customers to spend more time on the phone, the mobile provider comes up with a promotion whereby a customer spending three minutes on a call gets any time after that for free. The mean length of a random sample of 423 phone calls after the implementation of the promotion was found to be 7.33 minutes with a standard deviation of 1.25 minutes. Develop a 96% confidence interval for the difference between the mean length of a phone call before the promotion and the mean length of a phone call after the promotion.

Note that one-sided confidence intervals can also be constructed for the difference between two means.

ESTIMATING THE POPULATION PROPORTION
Confidence Interval for the Population Proportion

Suppose instead that we wish to estimate the proportion of the class of 100 fast-food buying university students who buy food at Top Lunch Café. This is an example of a population proportion – the proportion of the population that possesses a particular characteristic. Again, we have neither the time, resources, nor the inclination to investigate every single member of the class of 100 who buy lunch at Top Lunch Café. We therefore take a sample of 40 students. Of these 40 students, 30 buy their fast food from Top Lunch Café. Therefore, the proportion of this sample who buy from Top Lunch Café is $\frac{30}{40}$ = 0.75. This is an example of a sample proportion – the proportion of a sample that possesses a particular characteristic. In this case the 'characteristic' is 'buying lunch from Top Lunch Café'.

Like the sample mean for the population mean, the sample proportion can be used as a point estimate of the population proportion. However, as in the case of the sample mean, the point estimate approach is insufficient in the case of the population proportion. We therefore construct a confidence interval.

We have three variables to consider for the sample under consideration – the proportion $\hat{p}$ of students that eat at Top Lunch Café, the remaining proportion $1 - \hat{p}$ of students that do not eat at Top Lunch Café, and the size n of the sample. The proportion $1 - \hat{p}$ is denoted as $\hat{q}$ for simplicity. We then have:

$\hat{p} = 0.75$
$\hat{q} = 1 - \hat{p} = 1 - 0.75 = 0.25$
$n = 40$

Using these three variables, we can then define a Distribution of the Sample Proportion. As in the case of the sample mean, the theory suggests that if we were to take all 13.75 octillion samples of size 40, and measure the proportion of university students who eat at Top Lunch Café in each sample, we would find that these sample proportions are normally distributed with a mean equal to the population proportion p, and a variance equal to $\frac{\hat{p}\hat{q}}{n}$ (standard deviation = $\sqrt{\frac{\hat{p}\hat{q}}{n}}$). We can therefore define a random variable $\hat{p}$ which represents the proportion of a sample of 40 university fast-food buying students who eat at Top Lunch Café.

$\therefore \hat{p} \sim N(0.75, \frac{(0.75)(0.25)}{40})$
$= \hat{p} \sim N(0.75, 0.0047)$

The variance is 0.0047, therefore the standard deviation is $\sqrt{0.0047} = 0.0685$

We can therefore represent this distribution on a Normal Distribution curve, so that when we seek to construct a 95% confidence interval for the population proportion, what we are actually looking for is an interval between which 95% of the proportions of the samples of size 40 would lie. Represented graphically, what we seek is:

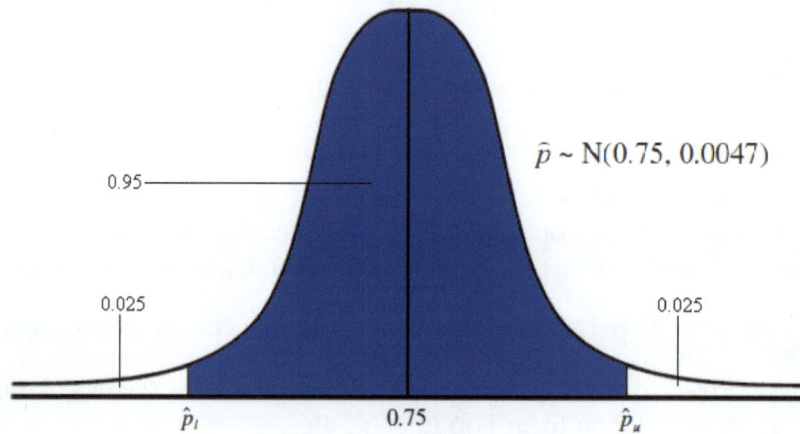

$\hat{p} \sim N(0.75, 0.0047)$

0.95

0.025   0.025

$\hat{p}_l$ 0.75 $\hat{p}_u$

We want two values $\hat{p}_L$ and $\hat{p}_U$ such that the probability that a sample proportion from a sample of size 40 would lie between these two values is 0.95. What this means is that the probability that the sample proportion from a sample of size 40 would lie outside this range is 0.05. From the graph, we can see the probability that the sample proportion would be less that the lower limit of the interval $\hat{p}_L$ is 0.025, and the probability that it would be greater than the upper limit of the interval $\hat{p}_U$ is also 0.025.

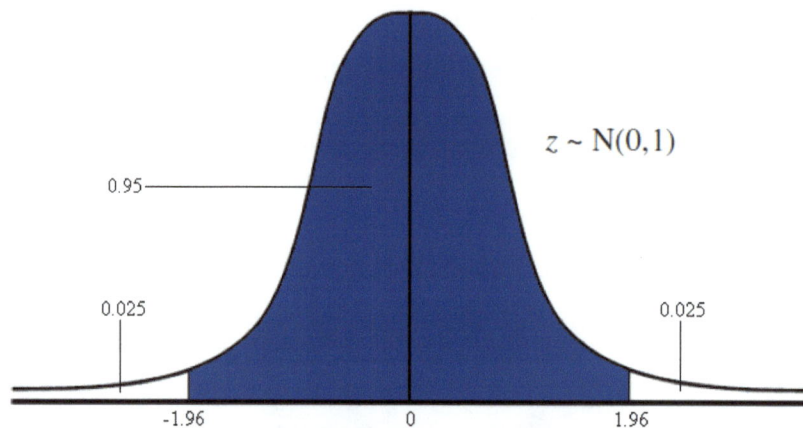

$z \sim N(0,1)$

0.95

0.025   0.025

-1.96   0   1.96

Just as we did for the sample mean, we use the z-curve to find the z-values that would correspond to probabilities of 0.025 in each of the left and right tails of the standard normal curve. The values that correspond to these areas are 1.96 for the right tail, and -1.96 for the left tail. We now need to de-standardize in order to find the values of the $\hat{p}$ variable that correspond to the z-values of -1.96 and 1.96. This would give us the lower and upper limits of the interval within which 95% of the means of random samples of size 40 would lie.
So:

$$-1.96 = \frac{\hat{p}_L - 0.75}{0.0685} \qquad\qquad 1.96 = \frac{\hat{p}_U - 0.75}{0.0685}$$

$$(-1.96)(0.0685) = \hat{p}_L - 0.75 \qquad\qquad (1.96)(0.0685) = \hat{p}_U - 0.75$$

$$-0.1343 = \hat{p}_L - 0.75 \qquad\qquad 0.1343 = \hat{p}_U - 0.75$$

$$0.75 - 0.1343 = \hat{p}_L \qquad\qquad 0.75 + 0.1343 = \hat{p}_U$$

$$\therefore \quad \hat{p}_L = 0.6157 \qquad\qquad\qquad \therefore \quad \hat{p}_U = 0.8843$$

Therefore, based on this particular sample the 95% confidence interval for the population proportion would be *0.6157≤ p ≤ 0.8843*. If somehow we were able to measure the proportion of students who buy lunch at Top Lunch Café for all 13.75 octillion possible samples of size 40 from the population of 100, 95% of those sample proportions would lie between 0.6157 and 0.8843. This means that 95% of all the possible point estimates for the population proportion based on samples of size 40 would lie between 0.6157 and 0.8843. We can therefore say that, based on all the possible point estimates from samples of size 40, the probability that the population proportion lies within the interval *0.6157≤ p ≤ 0.8843* is 0.95. This result is alternatively interpreted to mean that based on this sample of 40, we can say with 95% confidence that the population proportion will lie between 0.6157 and 0.8843.

Exercise 6.4

For each of the following scenarios, construct the required confidence interval for the population proportion.

(1) For a particular sample of size 67, $\hat{p} = 0.76$. Construct the following confidence intervals for the population proportion:

(a) 92%   (b) 90%   (c) 87%   (d) 96%   (e) 95%

(2) For a particular sample of size 43, $\hat{p} = 0.21$. Construct the following confidence intervals for the population proportion:

(a) 96%   (b) 91%   (c) 97%   (d) 82%   (e) 99%

(3) The quality assurance division of a cell phone manufacturing plant conduct tests a sample of 860 randomly selected phones. Eleven of these phones has a manufacturing defect. Use this particular sample to construct a 98% confidence interval for the proportion of defective cell phones manufactured by this plant.

(4) Rhona is cleaning her e-mail inbox. As part of the process, she selects 200 random e-mail messages and counts the number of e-mails that were forwarded to her. Of the 200 e-mails, 26% are 'forwards'. Use this particular sample of e-mails to develop a 95% confidence interval for the proportion of all Rhona's e-mails that are 'forwards'.

(5) A farmer on a pineapple farm in Mexico wishes to estimate the proportion of his pineapples that have been diseased as a result of a particularly severe case of an infection that has randomly attacked his pineapple plantation. He finds that 27% of a random sample of 117 pineapples have been afflicted with this mysterious infestation. Derive a 91% confidence interval for the proportion of pineapples in the entire plantation that have been infected.

(6) 15 out of a sample of 72 inmates from a particular prison population are left-handed. Develop a 96% confidence interval for the proportion of left-handed inmates in the entire prison population.

(7) Health inspectors examine a random sample of 64 sacks of flour at the neighborhood supermarket. 23 of these sacks of flour are found to be contaminated with weevils. Construct a 92% confidence for the proportion of the supermarket's entire stock of sacks of flour that are weevil-infested.

(8) Seven percent of a sample of 100 computer chips at a computer factory are found to be defective. Construct a 97% confidence interval for the proportion of all the factory's computer chips that are defective.

(9) A sample of 100 cows from a farm are tested. From this sample 19 cows are found to be infected with mad cow disease. Develop a 99% confidence interval for the proportion of the population of cows on that farm that are infected with mad cow disease.

(10) At a brick factory, the structural integrity of a sample of new-formed bricks is tested by dropping them from a pre-determined height. A brick fails the test by breaking apart. A sample of 153 bricks is thus tested, and 27 of them break apart. Develop a 94% confidence interval for the proportion of the factory's bricks that are structurally sound.

One-Sided Confidence Intervals for the Population Proportion
Upper Confidence Intervals

As was the case for the sample mean, suppose instead that we want to find a 95% confidence interval for the **minimum value** of the proportion of students from the class of 100 university students who buy food at Top Lunch Café.

We want only a minimum value above which 95% of the proportions for samples of size 40 would lie. Therefore the probability that the sample proportion is less than this minimum value would be 0.05 (1 − 0.95). Graphically, this situation would appear thus:

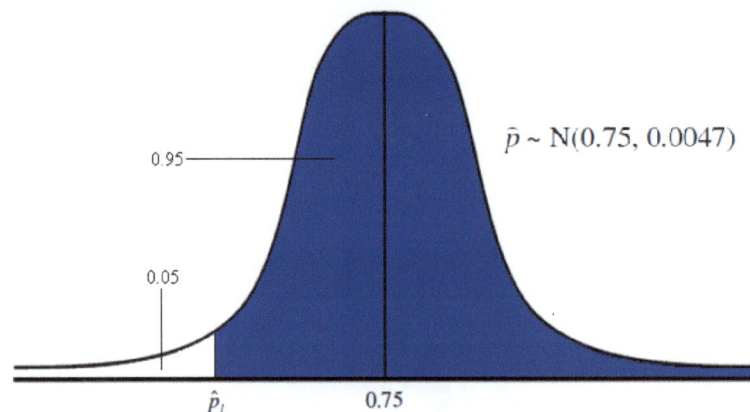

$$\bar{p} \sim N(0.75, 0.0047)$$

As in the case of the two-sided confidence intervals, we now go to the z-curve and look for the z-value that gives an area of 0.05 in the left tail of the standard normal curve:

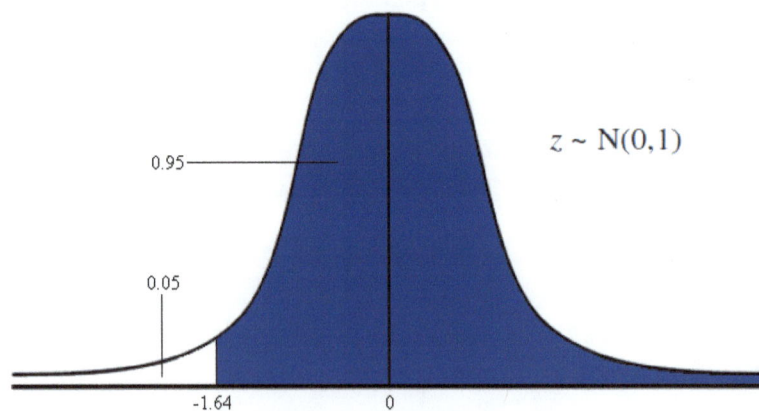

$$z \sim N(0,1)$$

From the z-table, this value is -1.64, so we need to de-standardize this z-value to find the value of $\overline{X}_L$ on the distribution of the sample mean that would correspond to the minimum value of our confidence interval.

$$-1.64 = \frac{\hat{p}_L - 0.75}{0.0685}$$

$$(-1.64)(0.0685) = \hat{p}_L - 0.75$$
$$-0.11234 = \hat{p}_L - 0.75$$
$$0.75 - 0.11234 = \hat{p}_L$$
$$\therefore \quad \hat{p}_L = 0.6377$$

So, we can say that the upper 95% confidence interval for the value of the proportion of students who eat at Top Lunch Café is: $p \geq 0.6377$. Based on this particular sample, we can say that if the proportions of all the possible samples of size 40 were tabulated, 95% of those sample proportions would be greater than 0.6377. This means that 95% of all the possible point estimators for the population proportion based on samples of size 40 would be greater than 0.6377. We can therefore say that, based on all the possible point estimators from samples of size 40, the probability that the population proportion lies within the interval $p \geq 0.6377$ is 0.95. This result is alternatively interpreted to mean that based on this sample of 40, we can say with 95% confidence that the population proportion would be greater than 0.6377.

Lower Confidence Intervals

Let us now reverse the situation. Suppose instead that we want to find a 95% confidence interval for the **maximum value** of the proportion of students from the class of 100 university students who buy food at Top Lunch Café. Again we use a sample of size 40. What we are in fact seeking is a value below which 95% of the possible point estimators lie. And we get this value based on our sample of size 40. So what we have graphically is:

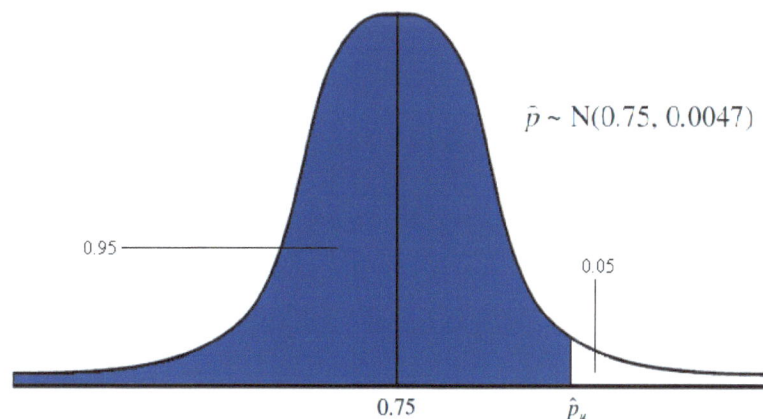

We now go to the z-table to find the standardized z-value that gives an area of 0.05 in the right tail of the z-curve. This value is 1.64:

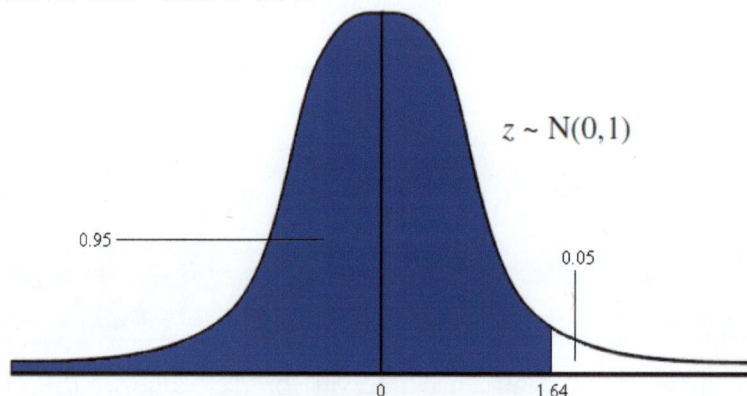

We need to de-standardize this z-value to find the value of $\hat{p}_u$ on the distribution of the sample proportion that would correspond to the maximum value of our confidence interval. So:

$$1.64 = \frac{\hat{p}_u - 0.75}{0.0685}$$

$$(1.64)(0.0685) = \hat{p}_u - 0.75$$

$$0.11234 = \hat{p}_u - 0.75$$

$$0.75 + 0.11234 = \hat{p}_u$$

$$\therefore \quad \hat{p}_u = 0.8623$$

We can therefore say that the lower 95% confidence interval for the value of the proportion of students who eat at Top Lunch Café is: $p \leq 0.8623$. Based on this particular sample, we can say that if the proportions of all the possible samples of size 40 were tabulated, 95% of those sample proportions would be less than 0.8623. This means that 95% of all the possible point estimators for the population proportion based on samples of size 40 would be less than 0.8623. We can therefore say that, based on all the possible point estimators from samples of size 40, the probability that the population proportion lies within the interval $p \leq 0.8623$ is 0.95. This result is alternatively interpreted to mean that based on this sample of 40, we can say with 95% confidence that the population proportion would be less than 0.8623.

Exercise 6.5
For each of the following scenarios, construct the required confidence interval for the population proportion.

(1) For a particular sample of size 67, $\hat{p} = 0.76$. Construct the following lower confidence intervals for maximum value of the population proportion p:
  (a) 92%   (b) 90%   (c) 87%   (d) 96%   (e) 95%

(2) For a particular sample of size 43, $\hat{p} = 0.21$. Construct the following upper confidence intervals for the minimum value of the population proportion p:
  (a) 96%   (b) 91%   (c) 97%   (d) 82%   (e) 99%

(3) The quality assurance division of a cell phone manufacturing plant conducts tests on a sample of 860 randomly selected phones. Eleven of these phones have a manufacturing defect. Use this particular sample to construct a 98% upper confidence interval for the minimum proportion of defective cell phones manufactured by this plant.

(4) Rhona is cleaning her e-mail inbox. As part of the process, she selects 200 random e-mail messages and counts the number of e-mails that were forwarded to her. Of the 200 e-mails, 26% are 'forwards'. Use this particular sample of e-mails to develop a 95% lower confidence interval for the maximum proportion of all Rhona's e-mails that are 'forwards'.

(5) A farmer on a pineapple farm in Mexico wishes to estimate the proportion of his pineapples that have been diseased as a result of a particularly severe case of an infection that has randomly attacked his pineapple plantation. He finds that 27% of a random sample of 117 pineapples has been afflicted with this mysterious infestation. Derive a 91% lower confidence interval for the maximum proportion of pineapples in the entire plantation that have been infected.

(6) 15 out of a sample of 72 inmates from a particular prison population are left-handed. Develop a 96% upper confidence interval for the minimum proportion of left-handed inmates in the entire prison population.

(7) Health inspectors examine a random sample of 64 sacks of flour at the neighborhood supermarket. 23 of these sacks of flour are found to be contaminated with weevils. Construct a 92% lower confidence for the maximum proportion of the supermarket's entire stock of sacks of flour that are weevil-infested.

(8) Seven percent of a sample of 100 computer chips at a computer factory is found to be defective. Construct a 97% upper confidence interval for the minimum proportion of all the factory's computer chips that are defective.

(9) A sample of 100 cows from a farm are tested. From this sample 19 cows are found to be infected with mad cow disease. Develop a 99% lower confidence interval for the maximum proportion of the population of cows on that farm that are infected with mad cow disease.

(10) At a brick factory, the structural integrity of a sample of new-formed bricks is tested by dropping them from a pre-determined height. A brick fails the test by breaking apart. A sample of 153 bricks is thus tested, and 27 of them break apart. Develop a 94% upper confidence interval for the minimum proportion of the factory's bricks that are structurally sound.

Confidence Interval for the Difference Between Two Population Proportions

Just as we did for population means, we can construct confidence intervals for the difference between two population proportions.

Example 6.2

Of a sample of 97 residents of Chaguanas, 36 were found to be smokers. A sample of 39 residents of the San Fernando contained 17 smokers. Construct a 96% confidence interval for the difference between the proportions of smokers in Chaguanas and San Fernando.

Solution

First things first. Find the proportion of smokers in each sample. For the Chaguanas sample, the proportion of smokers is $\frac{36}{97} = 0.37$. The proportion of smokers in the San Fernando sample is $\frac{17}{39} = 0.44$. The next step is to test each sample to see whether or not it conforms to the conditions that would allow use to define the distribution of the sample proportion. Recall from the definition of the Central Limit Theorem for Sample Proportions that these conditions are: $n\hat{p} > 5$, and $n\hat{q} > 5$.

So, for the Chaguanas sample: $n = 97$, $\hat{p} = 0.37$, $\hat{q} = 1 - 0.37 = 0.63$

$$\therefore n\hat{p} = (97)(0.37) \qquad n\hat{q} = (97)(0.63)$$
$$= 35.89 \qquad\qquad = 61.11$$

$n\hat{p}$ and $n\hat{q}$ are both greater than 5, so we are now free to construct the distribution of the sample proportion for the Chaguanas sample.

In the case of the San Fernando sample: $n = 39$, $\hat{p} = 0.44$, $\hat{q} = 1 - 0.44 = 0.56$

$$\therefore n\hat{p} = (39)(0.44) \qquad n\hat{q} = (39)(0.56)$$
$$= 35.89 \qquad\qquad = 61.11$$
$$= 17.16 \qquad\qquad = 21.84$$

$n\hat{p}$ and $n\hat{q}$ are both greater than 5, so we are now free to construct the distribution of the sample proportion for the San Fernando sample.

Our next step is to develop a Normal Distribution for the difference between the Port-of-Spain and San Fernando proportions. We must first develop separate distributions for the Chaguanas and San Fernando proportions.

As always, we first define the random variables.

Let $\hat{c}$ be the random variable 'proportion of a sample of 97 Chaguanas residents who smoke'.

$\hat{p} = 0.37$, $\hat{q} = 0.63$, $n = 97$

$$\therefore \hat{c} \sim N(0.37, \frac{(0.37)(0.63)}{97}) \Rightarrow \hat{c} \sim N(0.37, 0.0024)$$

Let $\hat{s}$ be the random variable 'proportion of a sample of 39 San Fernando residents who smoke'.

$\hat{p} = 0.44$, $\hat{q} = 0.56$, $n = 39$

$$\therefore \hat{s} \sim N(0.44, \frac{(0.44)(0.56)}{39}) \Rightarrow \hat{s} \sim N(0.44, 0.0063)$$

We now proceed to develop the Normal Distribution for the difference between the proportion of smokers from the Chaguanas sample and the proportion of smokers from the San Fernando sample.

Let $\hat{s} - \hat{c}$ be the random variable 'the difference between the proportion of smokers from the San Fernando sample and the proportion of smokers from the Chaguanas sample.

$$E(\hat{s} - \hat{c}) = E(\hat{s}) - E(\hat{c})$$
$$= 0.44 - 0.37$$
$$= 0.07$$

$$Var(\hat{s} - \hat{c}) = Var(\hat{s}) + Var(\hat{c})$$
$$= 0.0063 + 0.0024$$
$$= 0.0087$$

$$\therefore \hat{s} - \hat{c} \sim N(0.07, 0.0024)$$

We let $d = \hat{s} - \hat{c}$
Therefore, $d \sim N(0.07, 0.0024)$

We can represent the distribution for d on a normal curve, so that when we seek to construct a 96% confidence interval, what we are actually looking for is an interval between which 96% of the differences between the means of European and Hollywood films would lie.

We now proceed with the construction of the confidence interval as we did previously. First, consider the normal distribution for the difference between the proportions of San Fernando and Chaguanas smokers from the respective samples.

The 96% confidence interval from this distribution would give the following situation:

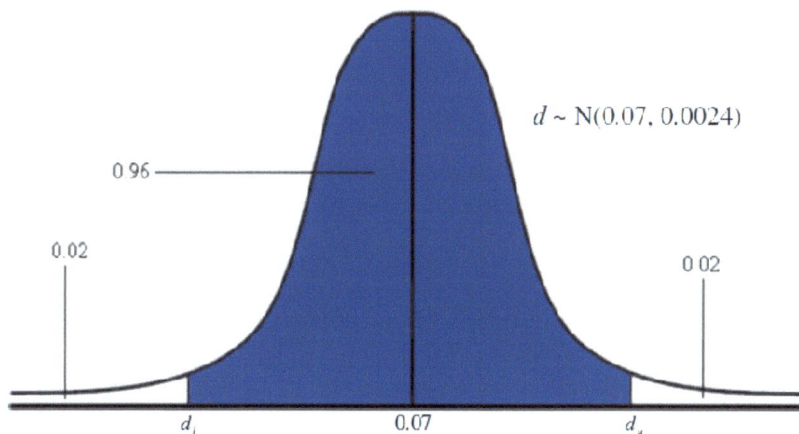

$d \sim N(0.07, 0.0024)$

0.96

0.02

0.02

d_i 0.07 d_4

We therefore seek two values d_L and d_U such that the probability that getting a value between these values is 0.96. We can also say that 96% of the values for the difference between the proportion of San Fernando and Chaguanas lie between these two values.

As usual, we go to the z-table to find the z-values that would give us probabilities of 0.02 at both extremities of the Standard Normal Distribution Curve.

Those values are -2.05 and 2.05 respectively.

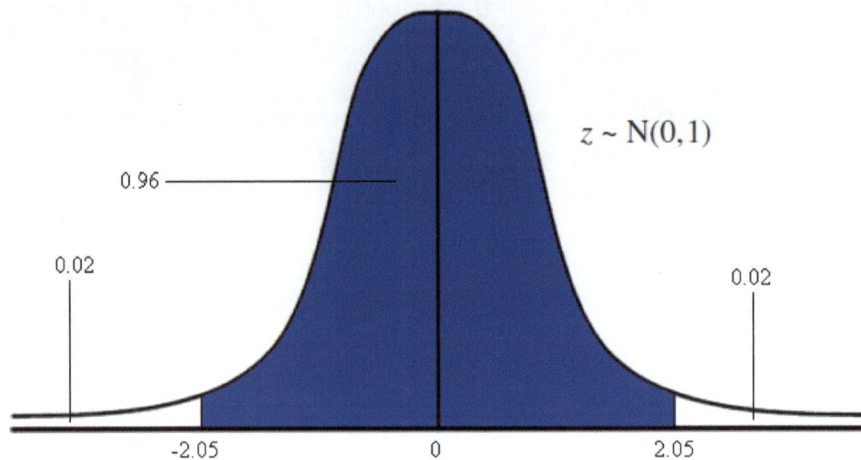

Therefore, we can now say that:

$$-2.05 = \frac{d_L - 0.07}{\sqrt{0.0024}}$$

$$= \frac{d_L - 0.07}{0.049}$$

$$(-2.05)(0.049) = d_L - 0.07$$

$$-0.1005 = d_L - 0.07$$

$$0.07 - 0.1005 = d_L$$

$$-0.0305 = d_L$$

$$\therefore \quad d_L = -0.0305$$

$$2.05 = \frac{d_U - 0.07}{\sqrt{0.0024}}$$

$$= \frac{d_U - 0.07}{0.049}$$

$$(2.05)(0.049) = d_U - 0.07$$

$$0.1005 = d_U - 0.07$$

$$0.07 + 0.1005 = d_U$$

$$0.1705 = d_U$$

$$\therefore \quad d_U = 0.1705$$

Therefore, the 96% confidence interval for the difference between the proportion of smokers in the San Fernando and Chaguanas populations is (-0.0305, 0.1705). The difference between the proportions of smokers in the San Fernando and Chaguanas populations ranges from one extreme of -0.0305 to the other extreme of 0.1705. What is the significance of -0.0305? Remember, the variable is 'the difference between the proportions of smokers in samples from San Fernando and Chaguanas. This variable can be defined either as $\hat{s} - \hat{p}$ or as $\hat{p} - \hat{s}$, since all we are concerned with is the **difference**. We chose $\hat{s} - \hat{p}$ simply because this is the configuration that would give a positive value for the difference, because $\hat{s}$ has a greater mean than $\hat{p}$.

Using $\hat{p} - \hat{s}$ would give us a mean of -0.07 for the mean difference between the proportions of smokers from samples of the Chaguanas and San Fernando populations. From a strict mathematical standpoint, we can choose either scenario and reach the same conclusion. However, using a negative mean would complicate calculations unnecessarily, so as a general rule, anytime we have to develop a normal distribution for the difference between two means or the difference between two proportions, we utilize the configuration that would give a positive value for the mean difference.

Having chosen the $\hat{s} - \hat{p}$ configuration, a value of -0.0305 simply means that $\hat{s}$ is ***less than*** $\hat{p}$ by 0.0305. The value of 0.1705 means that $\hat{s}$ is greater than $\hat{p}$ by 0.1705.

Exercise 6.6
For each of the following situations, construct the requested confidence interval for the difference between two population proportions:

(1) Half of a sample of 66 North Koreans believe that their country is stockpiling nuclear weapons. Three-fifths of a sample of 75 South Koreans believe that North Korea is stockpiling nuclear weapons. Construct a 95% confidence interval for the difference between the proportion of South Koreans who believe that North Korea is stockpiling nuclear weapons and the proportion of North Koreans who hold that view.

(2) In the Democratic primary in the 2008 U.S.A Presidential Elections, 197 of a random sample of 235 voters in South Carolina supported the Independent candidate. A similar sample of 198 voters in North Carolina contained 101 supporters of the Independent candidate. Use this data to construct a 96% confidence interval for the difference between the proportion of 'Independent' supporters in South Carolina and the proportion of 'Independent' supporters in North Carolina.

(3) Seven percent of a random sample of 100 religious e-mails is forwarded, while 79 percent of a similar sample of 100 pornographic e-mails is forwarded. Develop a 92% confidence interval for the difference between the proportion of forwarded pornographic e-mails and the proportion of forwarded religious e-mails.

(4) In the Caribbean island nation of Cuba, 90 percent of a sample of 95 residents of Havana is found to be users of fluorescent bulbs. 67 percent of a sample of 87 residents of Santiago de Cuba is found to be users of fluorescent bulbs. Develop a 99% confidence interval for the difference between the proportion of flourescent bulb users in Havana and the proportion of fluorescent bulb users in Santiago de Cuba.

(5) A sample of 232 records for women from the records office in London indicate that 75% of married London women got married before the age of 30. A similar sample of 212 records for men indicates that 63 percent of men get married before the age of 30. Develop a 94% confidence interval for the difference between the proportion of London men who marry before age 30 and the proportion of London women who marry before age 30.

(6) At a DVD factory in China, 60 of a random sample of 1,000 DVDs contained structural defects. A quality assurance procedure was implemented, and a sample of 1,500 DVDS contained 60 defects. Construct a 90% confidence interval for the difference between the proportion of defective DVDs before the implementation of the quality assurance procedure and the proportion of defective DVDs after the implementation of the quality assurance procedure.

(7) At a poultry farm in Mausica, Trinidad, a sample of 300 chicken eggs contain 41 bad eggs. A sample of 450 duck eggs contains 77 bad eggs. Develop a 93% confidence interval for the difference between the proportion of bad chicken eggs and the proportion of bad duck eggs.

(8) Soil analysts are testing two different types of soil – 'soil A' and 'soil B'. Part of the test involves planting a particular seed X in both types of soil and monitoring the progress of the plants that grow from these seeds. 16 of a random sample of 96 seeds from soil A do not survive, while 18 of a random sample of 74 seeds from soil B don not survive. Develop a 96% confidence interval for the difference between the proportion of seeds that survive in soil A and the proportion of seeds that survive in soil B.

(9) Before the onset of the global financial recession, fifteen percent of a random sample of 95 army privates enrolled in a financial literacy program offered by the military. With the recession, a random sample of 80 army privates revealed that 65 of them had enrolled in the program. Based on these figures, develop a 94% confidence interval for the difference between the proportion of army privates enrolling in the financial literacy program before the recession and the proportion of army privates enrolling in the program after the recession.

(10) A telecommunications monitoring firm is comparing two makes of mobile phones – Phone P and Phone Q. A random sample of 56 buyers of phone X was tracked, and 23 of them had complaints about the operation of their phone within the first six months of ownership. A similar exercise was carried out with a sample of 78 buyers of Phone Q, and 38 of them registered complaints about the operation of their phone within the first six months of ownership. Use these figures to construct a 96% confidence interval for the difference between the proportion of Phone P users and the proportion of Phone Q users who record complaints about their phones within the first six months of ownership.

7

Hypothesis Testing

The central premise of hypothesis testing using the Normal Distribution is the use of one or two sample statistics to investigate *significance of difference*. For example, we can investigate whether or not the new value of a population parameter as estimated by a sample statistic is significantly different from the old value. If the difference between the new value and the old value is significant, then we say that the value of the old population parameter has *changed*. This change may be expressed as an *increase*, a *decrease*, or just simply, a *change*, where we are concerned only with the fact that a change has in fact occurred, without being worried about the precise nature of that change. If the difference between the old value and the new value is not significant, we can conclude that the value of the old population parameter has not changed. We can also investigate whether the difference between the values of a given population parameter for two independent populations is significant. If this difference is significant, we say that the two values of the given population parameter for the two populations are *different*, and we may go further to say that one value is *greater than* the other value, or alternatively that one value is *less than* the other value. If the difference is not significant, we say that the two values are *not different,* and for most practical intents and purposes, the values are considered to be the same.

There are two hypotheses under consideration – the **null hypothesis**, and the **alternative hypothesis**. The null hypothesis is the one that asserts **no significant difference**, while the alternative hypothesis suggests **a significant difference**.

Hypothesis Testing using the Normal Distribution involves two hypotheses – the null hypothesis, written as H_0, and the alternative hypothesis, written as H_1.

The null hypothesis suggests that there is 'no significant difference'; while the alternative hypothesis suggests that there is in fact 'a significant difference'.

For example, we may know from previous data that the mean amount of money spent by a student from our university class of 100 students who buy food at Top Lunch Café is $25. We then take a sample of 40 such students, measure the mean of the sample, and come up with a sample mean of 27. We may therefore wish to investigate whether or not the result from our sample points to a mean that is now not equal to 25, or more specifically whether or not it points to a mean that is now greater than 25. In other words, we are testing to see if the difference between our sample mean of 27 and the original population mean of 25 is significant. If this difference is found to be significant, then we say that the population mean as estimated by our sample of 40 is different to the original population mean. This implies that the population mean has changed.

Similarly, we may know from past data that of our class of 100 university students who purchase food at Top Lunch Café, the proportion who buys Trinidad and Tobago cuisine is 0.75. However, when we take a sample of 40 students, we find that 25 of them buy Trinidad and Tobago cuisine, giving a proportion of $\frac{25}{40} = 0.625$. We therefore may wish to test whether this figure obtained suggests that the proportion of our population who buys Trinidad and Tobago cuisine is now not equal to 0.75, or more specifically whether this figure means that this proportion is now less than 0.75. In other words, if the difference between our sample proportion of 0.625 and the original population proportion of 0.75 is found to be significant, then we say that the population proportion as estimated by our sample of size 40 is different from the original population proportion. Alternatively, we can get more specific and say that the old population proportion has decreased. We will here consider only hypothesis tests done using large samples.

How 'Different' is 'Significant'?

All of this naturally begs the question – What is the threshold point beyond which a 'difference' becomes 'significant'? There is no 'fixed' answer to this question – in fact the person doing the hypothesis test can usually choose where that threshold point lies. This is done by selecting the 'significance level' of the hypothesis test.

The significance level of a hypothesis test is the probability that a 'difference' in the test under consideration would be 'significant'. The lower the level of significance, the less likely it is that the 'difference' under consideration would be 'significant', hence significant differences at lower significance levels usually attract more attention than significant differences at higher significance levels.

TESTING THE POPULATION MEAN
Two-Tailed Hypothesis Tests

Consider once more our class of 100 university students who spend money buying fast food at Top Lunch Café. We know from past data that the mean expenditure on fast food in this class is $25. We know also that the standard deviation of this expenditure is $5. We take a sample of size 40 and calculate a mean expenditure of $27.

Suppose we wish to test whether our sample mean of 27 is an indication that the mean is now different from 25. Recall in chapter 6 that we had calculated a 95% confidence interval for the population mean using a sample of size 40, a sample mean of 25, and a population standard deviation of 5. If our sample mean of 27 lies within this confidence interval, then we say that it is not significantly different from 25. We conclude then that our sample mean of 27 does not indicate any change in the population mean from the original value of 25. If our value of 27 falls outside the confidence interval, then we say that it is significantly different from 25, and we therefore can say that sample mean of 27 is indeed an indication that there has been a change from the original value of 25.

Because we are testing to see if the value 27 lies within a 95% confidence interval, that leaves us with a region of 5% outside of this interval. We are simply testing a difference without being specific as to the precise nature of that difference, so in addition to the case where the test value is greater than the original value, we must also consider the case where the value being tested is less than the original value. This means that a test value may either be 'significantly less than' or 'significantly greater than' the original value of 25. The 95% confidence interval would therefore be a two-sided confidence interval, implying a remaining region of 5% equally divided into two regions of 2.5% in either tail of the Normal Distribution Curve. Both regions of 2.5% together comprise the **region of significance**, giving us a hypothesis test at the 5% significance level (2 x 2.5%). The 95% confidence interval is therefore the **region of non-significance**.

The region of non-significance is also called the **non-rejection region**, since a test value and its associated test statistic falling in this region implies a difference that is 'not significant', and hence in that case we *do not reject* the null hypothesis. The region of significance is also called the **rejection region**, because once the test value and its associated test statistic falls within this region, the implied 'significant difference' means that we must *reject* the null hypothesis, accepting the alternative hypothesis as a result.

The null hypothesis (H_0) for our test under consideration is that there is no change in the previous population mean ($\mu = 25$). The alternative hypothesis (H_1) would be that there has been a change in the population means ($\mu \neq 25$). Written in the style and jargon of hypothesis testing, we would have:

H_0: $\mu = 25$
H_1: $\mu \neq 25$

Recall from chapter 6, that our 95% confidence interval for the population mean using a sample of size 40, a sample mean of 25, and a population standard deviation of 5 was: *23.45 $\leq \mu \leq$ 26.55*. If the value of our sample mean falls outside of this interval, then we can say that at the 5% significance level, our sample mean is 'significantly different' from the original population mean of 25. If the value of the sample mean falls within the interval, then we say that the sample mean is 'not significantly different' from the original population mean of 25. Graphically we have:

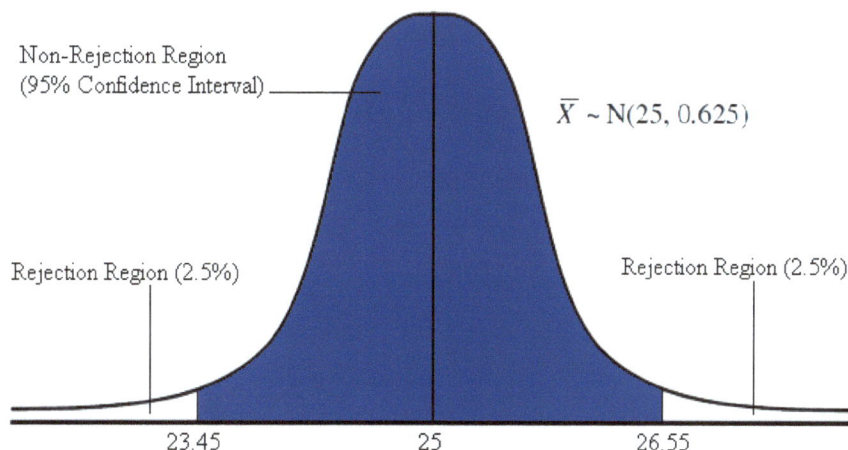

Non-Rejection Region
(95% Confidence Interval)

$\bar{X} \sim N(25, 0.625)$

Rejection Region (2.5%)

Rejection Region (2.5%)

23.45   25   26.55

On this distribution, it is clear that our sample mean of 27 will fall outside of the 95% confidence interval, as seen in this graph:

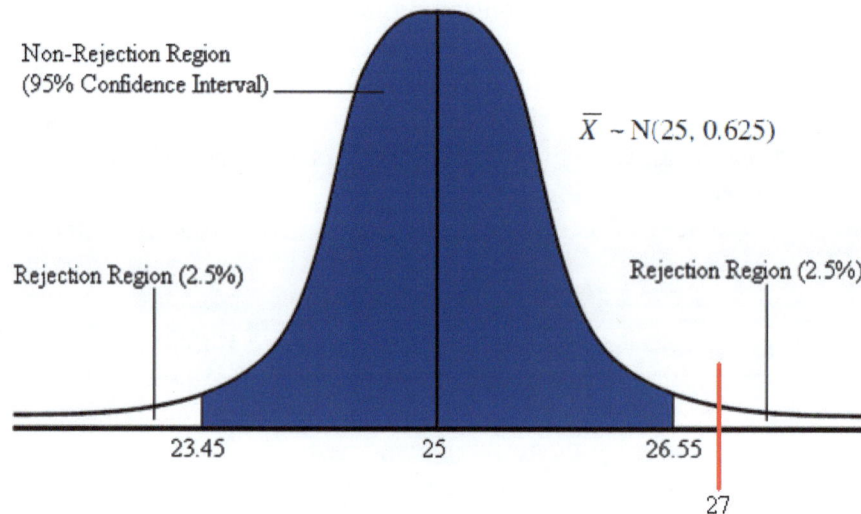

We therefore say that the value 27 is 'significantly different' from the original population mean of 25, in which case we say that the population mean has changed. We therefore reject the null hypothesis and accept the alternative hypothesis, in this case concluding that based on the evidence of our sample mean, we can say that the original population mean has changed and is not now equal to 25.

The great thing about doing these hypothesis tests is that we do not need to have prior knowledge of the corresponding confidence interval, neither do we need to know the relevant sampling distribution from which the corresponding confidence interval was derived. We can derive the real or implied sampling distribution, and we do not need to know the actual limits of the confidence interval that corresponds to the hypothesis test that we are doing.

The first step would be to derive the real or implied distribution of the sample mean from which the corresponding confidence interval was derived. We do this in the following way:

(a) *Let the mean of this distribution be equal to the value of the null hypothesis*
(b) *Construct the variance of this distribution using the appropriate population parameters or sample statistics.*

The resulting distribution is called **The Sampling Distribution of the Null Hypothesis**. This is a concept that explicitly links the hypothesis test to its corresponding confidence interval by using the available population parameters and sample statistics to derive the sampling distribution that would have been used to construct the confidence interval that would give us the non-rejection region of the hypothesis test and the resultant rejection region. The test value of the hypothesis test is then standardized using this distribution, giving us the test statistic, which can then be used in the Critical Value Method or the P-Value Method to carry out the hypothesis test.

So in this case, following (a), the mean of the Sampling Distribution of the Null Hypothesis will be equal to 25, since this is the value of the null hypothesis (**H_0: μ = 25**). Following (b), we use the population variance of 5^2 in tandem with the sample size of 40 to derive the variance of the sampling distribution as $\dfrac{5^2}{40}$. So the Sampling Distribution of the Null Hypothesis in this case has a mean of 25, and a variance of $\dfrac{5^2}{40}$.

So, $H_0 \sim N(25, \dfrac{5^2}{40}) \Rightarrow H_0 \sim N(25, 0.625)$

Using this distribution, we now standardize the value of our sample mean (27) to find the **Test Statistic**, which we will then place on the Standard Normal Distribution to see where it lies in relation to the **Critical Values**, which are the z-values on the Standard Normal Distribution which would give us the required areas in the region of significance or the rejection region.

A two-sided hypothesis test at the 5% significance level means that we would have a rejection region of 2.5% in each of the extremities of the Standard Normal Distribution curve. The z-values that will give us an area of 0.025 in each extremity would be -1.96 and 1.96. These are the critical values.

The test statistic is calculated as follows: $z_{27} = \dfrac{27 - 25}{\sqrt{0.625}} = \dfrac{27 - 25}{0.7906} = 2.53$

On a graph we would have the following situation:

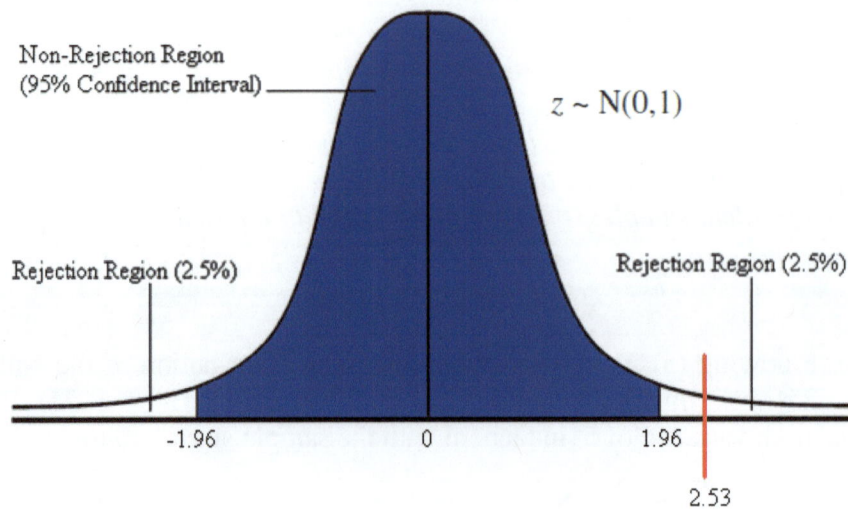

It is clear here that our test statistic falls within the rejection region. We therefore reject the null hypothesis, concluding in the process that the original population mean has changed. In other words, the population from which our sample of size 40 was drawn has a mean that is different from 25.

The method of performing hypothesis tests where we compare the test statistic to the critical value is called the **Critical Value Method**.

The P-Value Method

The P-Value Method is an alternative method of performing hypothesis tests which gives the same result as the Critical Value Method. The Critical Value Method and the P-Value Method in hard practical terms represent the flip side of the same coin. At this point, if it is necessary to go back to chapter 3 to revise exactly what the p-value is, then do so, as your knowledge and understanding of the p-value will be assumed from this point.

Let us go through the procedure for the p-value method using the example we have just worked. We already have the critical values (-1.96 and 1.96), and the test statistic (2.53). A two-tailed hypothesis test at the 5% level of significance gives us an area of $\frac{0.05}{2} = 0.025$ in the extremity of each tail. The area in the rejection region that corresponds to the significance level of the hypothesis test is referred to as 'alpha' (Greek symbol α). In this case where our significance level is 5%, our α would therefore be equal to 0.05. However, since this is a two-tailed test we will be concerned with the area in each tail, which would be equal to: $\frac{\alpha}{2} =$

$\frac{0.05}{2} = 0.025$

Now, find the p-value for 2.53. This is 0.0057. We now compare this p-value of 0.0057 to the $\frac{\alpha}{2}$ value of 0.025:

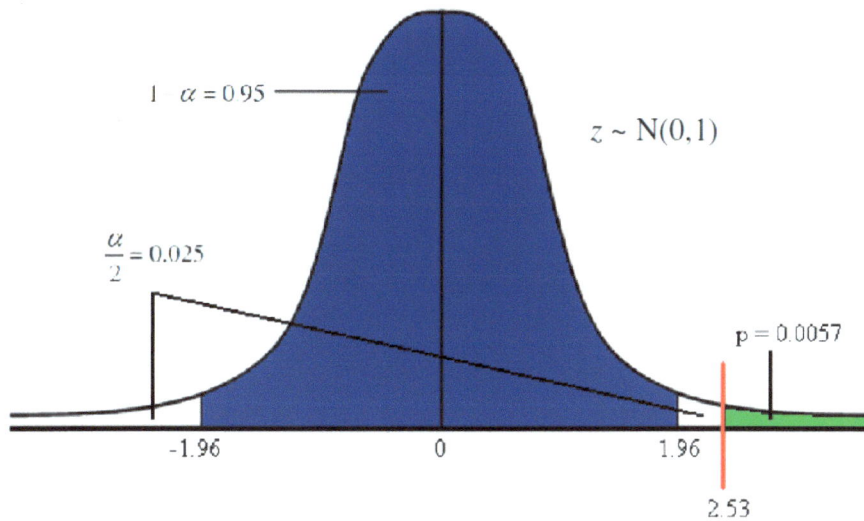

From this graph, we can clearly see that the p-value of the test statistic 2.53 (0.0057) is less than the $\frac{\alpha}{2}$ value (0.025), which is also the p-value of the critical value (1.96). If the test statistic were anywhere in the non-rejection region (shaded in blue), it is plain to see that the associated p-value would then be greater than the $\frac{\alpha}{2}$ value. This allows us to now enunciate the method by which we arrive at conclusions for hypothesis tests using the p-value method.

For a two-tailed hypothesis test at the A% level of significance:

(i) If $p < \frac{\alpha}{2}$, we reject the null hypothesis H_0

(ii) If $p \geq \frac{\alpha}{2}$, we do not reject the null hypothesis H_0

For a one-tailed hypothesis test at the A% level of significance:

(i) If $p < \alpha$, we reject the null hypothesis H_0

(ii) If $p \geq \alpha$, we do not reject the null hypothesis H_0

Note: $\alpha = \frac{A}{100}$

When p = α

The situation when the p-value is equal to the α value or the $\frac{\alpha}{2}$ value is one where the test statistic is equal to the critical value. In this case we do not reject the null hypothesis. The reason for this is that confidence intervals are defined in a manner that includes the actual limits of the interval as part of the interval itself. The critical value would therefore be contained within the corresponding confidence interval that comprises the non-rejection region. Therefore when the test statistic is equal to the critical value, it forms part of the non-rejection region, and hence we do not reject the null hypothesis H_0.

When p = 0

Very often in hypothesis tests, we encounter situations where the p-value of a particular test statistic is equal to zero. This happens when the test statistic is very large, so large in fact that it very literally is 'off the charts'. This large value of the test statistic is so far inside the rejection region that the area more extreme than the test statistic is *very small*, so very small that it can be approximated to zero for most applications with no loss of accuracy. In cases like this, we *always* reject the null hypothesis, since an effective p-value of zero will always be less than any value of α or $\frac{\alpha}{2}$ that we can think of.

Procedure for carrying out Hypothesis Tests (Critical Value Method)

The Critical Value Method for carrying out hypothesis tests involving the Normal Distribution can be summarized as follows:

(a) *Determine the Null and the Alternative Hypotheses*
(b) *Derive the Sampling Distribution of the Null Hypothesis*
(c) *Find the standardized z-score of the test value using the distribution from (b). This is the test statistic*
(d) *Determine the critical value(s) on the Standard Normal Distribution using the area(s) in the rejection region.*
(e) *Place the test statistic on the Standard Normal Distribution and locate it relative to the critical value(s)*
(f) *Draw the appropriate conclusion*

Procedure for carrying out Hypothesis Tests (P-Value Method)

The P-Value Method for carrying out hypothesis tests involving the Normal Distribution can be summarized as follows:

(a) *Determine the Null and the Alternative Hypotheses*
(b) *Derive the Sampling Distribution of the Null Hypothesis*
(c) *Find the standardized z-score of the test value using the distribution from (b). This is the test statistic*
(d) *Determine the p-value for the test statistic found in (c)*
(e) *Compare the p-value from (d) to α (one-tailed test) or $\frac{\alpha}{2}$ (two tailed test)*
(f) *Draw the appropriate conclusion*

Exercise 7.1

For the sampling distribution of the null hypothesis $H_0 \sim N(25, 0.625)$, and using the test values listed below, use both the critical value method and the p-value method to perform the following hypothesis test at the 5% significance level:

H_0: μ = 25
H_1: μ ≠ 25

Use the following values of $\overline{X}$ to calculate the test statistic:
(a) 22.78   **(b)** 24.32   **(c)** 20.01   **(d)** 26.77   **(e)** 28.54   **(f)** 26.03

Exercise 7.2

(1) Fifty years ago, the average height of the trees in a forest in north-east Trinidad was found to be 4,563 cm. The standard deviation of the heights of these trees at that time was 616 cm. Last week, a group of foresters measured the heights of 75 randomly selected trees from this forest and found that the average height of these trees was 4,398 cm. Using this available information, perform a hypothesis test at the 4% significance level to investigate whether or not the mean height of the trees in this forest has changed.

(2) In 1956, the mean number of children born to an American woman during her reproductive years was normally distributed with a mean of 3.91 and a standard deviation of 1.15. In 2006, the mean number of children born to a random sample of 63 American women was found to be 2.09 with a standard deviation of 0.95. Perform a hypothesis test at the 3% level of significance to determine if there has been a change in the average number of children born to an American woman between the years 1956 and 2006.

(3) After many years of medical practice Dr. Jones can say with confidence that the average glucose level of a patient taking a medical exam at his office is about 6 mmol/l. A junior doctor decides to test this claim using statistics theory. He chooses 60 patients at random and measures their blood glucose levels. The measurements yield an average glucose level of 6.45 mmol/l with a standard deviation of 1.93 mmol/l. Test at the 7% significance level the veracity of Dr. Jones' claim.

(4) The manufacturer of a new brand of intercontinental ballistic missile states that its missiles have an average range of 6,000 miles. As part of a series of missile tests to verify this claim, the rulers of a certain country that is interested in buying these missiles test fire 100 of them. The average range of these missiles is measured to be 5,978 miles with a standard deviation of 125 miles. At the 6% significance level, can we justify the claim of the manufacturer?

(5) A supermarket owner believes that the customers at his grocery normally spend about $15,000 each week on dairy purchases. In an attempt to test this, he has spent all 52 weeks of the last year monitoring the weekly dairy sales. His analysis revealed that the mean amount of money spent weekly on dairy products for the last year was $17,285.73, with a standard deviation of $6,060.80. On the basis of a hypothesis test at the 2% significance level, can the supermarket owner's belief be considered to be true?

(6) The residents of a rural village in Toco swear that they usually spend about 30 minutes waiting for the second bus into town after the first one has gone. The waiting times between the first bus and the second bus are measured on 56 occasions, and those waiting times are found to have a mean of 32 minutes and a standard deviation of 7.7 minutes. Can a hypothesis test at the 10% significance level justify the residents' claim?

(7) The Environmental Management Authority in Trinidad and Tobago estimates that the typical noise level at a soca party is approximately 180 decibels. One year, a group of statistics students measure the decibel levels at 40 soca parties. The average decibel level at these parties was measured to be 165 decibels, with a standard deviation of 53 decibels. Using this information, perform a hypothesis test at the 7% significance level and use the results to form a conclusion about the Authority's claim.

(8) Most fans of a soccer league in Spain believe that the amount of extra time added on to each game is about 3 minutes. 45 of these games from a particular season are analyzed, and the average amount of extra time is found to be 2.5 minutes with a standard deviation of 2.0 minutes. Use the results of a hypothesis test at the 8% significance level to determine the truth of the fans' belief.

(9) A small suburban radio station claims to play an average of 40 soft rock songs every day. A passionate soft rock listener seeks to investigate this claim by examining the station logs for 75 random days. The average number of soft rock songs played over these 75 days is 36.8, and the standard deviation is 7.67. What would be the listener's conclusion if she investigates this claim using a hypothesis test at the 9% level of significance?

(10) The Quality Assurance technicians at a soft drink factory in South Oropouche takes samples of soft drink dispensed by the two-litre machine which automatically dispenses two litres of soft drink into the empty bottles. If the average amount of soft-drink deviates significantly from two litres, adjustments are carried out to the dispensing machine. The latest sample of 100 bottles yields an average of 1.97 litres dispensed, with a standard deviation of 0.03 litres. A hypothesis test at the 5% significance level is part of the methodology employed by the Quality Team. Does this particular sample suggest that repairs to the two-litre machine are due?

One-Tailed Hypothesis Tests

Suppose we wanted to get more specific than simply testing for a 'difference', or a 'change'? Let's say for example that we want to test for a 'negative change', a 'negative difference', or a 'decrease'. Alternatively, we may want to test for a 'positive change', a 'positive difference, or an 'increase'.

In the case of the university class of 100 students, in all of these scenarios, the null hypothesis would still be H_0: $\mu = 25$. However, the alternative hypothesis would depend on the exact type of difference or change that we are testing for. In testing for a negative difference, we use a *left-tailed test*, and in testing for a positive difference, we employ the *right-tailed test.*

Left-Tailed Hypothesis Test

Consider again our present example. It is known from past records that the mean amount of money spent buying lunch by the students in our university class of 100 is $25. Suppose that our sample of 40 students yielded a mean of $23. Can we then conclude based on this sample that the mean amount of money spent on lunch has actually decreased? We perform a left-tailed hypothesis test at the 5% level of significance. The hypotheses will be formulated as follows:

H_0: $\mu = 25$
H_1: $\mu < 25$

Performing a left-tailed hypothesis test at the 5% significance level is the same as testing the value under consideration to see if it lies within the 95% upper confidence interval. Recall from chapter 6 that the 95% upper confidence interval where the sample mean was 25, the sample size was 40, and the population standard deviation was known to be 5 was: $\mu \geq 23.695$.

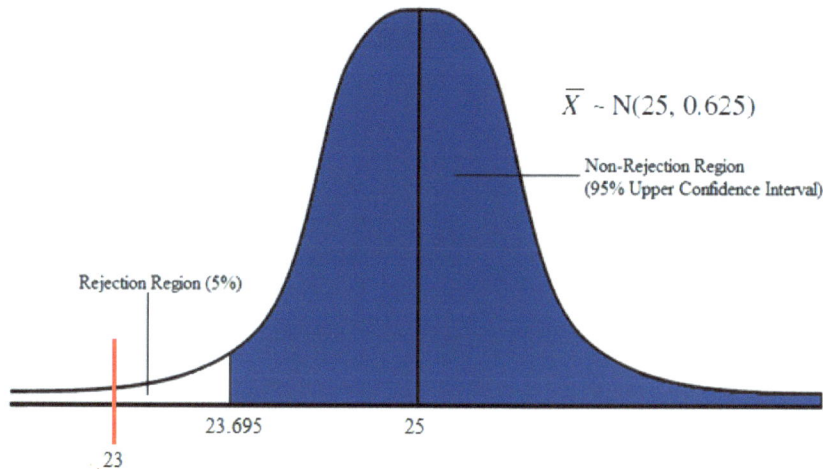

$\overline{X} \sim N(25, 0.625)$

Non-Rejection Region
(95% Upper Confidence Interval)

Rejection Region (5%)

23.695   25

.23

It is plain to see that the value 23 falls outside of the 95% upper confidence interval, so we can therefore regard the difference between the sample mean of 23 and the original population mean of 25 as a 'significant difference'. We therefore reject the null hypothesis and accept the alternative hypothesis, concluding that the mean has in fact decreased.

As in the case of the two-tailed hypothesis test, we do not need the actual confidence interval in order to perform the hypothesis test. We can derive the Sampling Distribution of the Null Hypothesis, and then find the appropriate test statistic, which we will then place on the standard normal curve and use its location there to guide us to a solution.

The mean of the Sampling Distribution of the Null Hypothesis will be 25, which is the value of the null hypothesis (H_0: $\mu = 25$). The variance of this distribution will be equal to the known population variance (5^2), divided by the sample size (40).

So, $H_0 \sim N(25, \dfrac{5^2}{40}) \Rightarrow H_0 \sim N(25, 0.625)$

We then use this distribution to find the standardized z-value that would correspond to the test value of 23: This would be: $z_{23} = \dfrac{23-25}{\sqrt{0.625}} = \dfrac{23-25}{0.7906} = -2.53$.

The z-score on the standard normal curve that would give an area of 0.05 in the left tail is -1.64. Graphically, we now have:

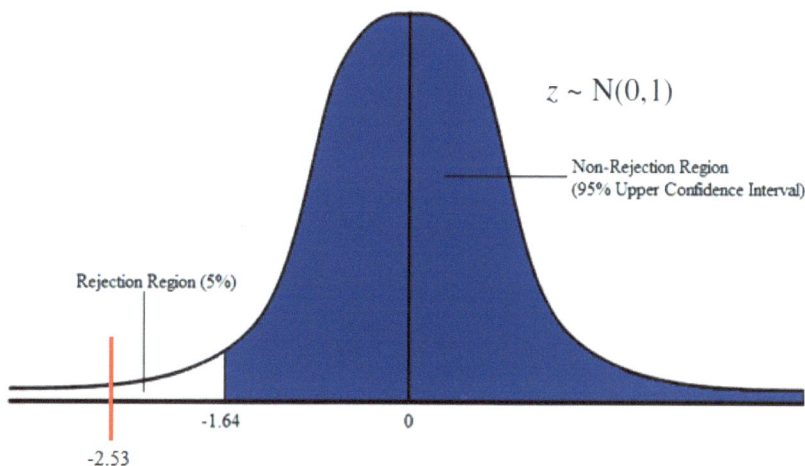

$z \sim N(0,1)$

Non-Rejection Region
(95% Upper Confidence Interval)

Rejection Region (5%)

-1.64   0

-2.53

117

It is clear that the test statistic falls outside of the non-rejection region shaded in blue (the 95% upper confidence interval). Therefore we reject the null hypothesis and conclude that the new population mean as estimated by our sample mean of 23 is in fact less than the original population mean of 25.

As an additional exercise, do this example using the p-value method. Verify that $p = 0.0057$, which is clearly less than the α value of 0.05. We therefore reject the null hypothesis as above.

Exercise 7.3
For the sampling distribution of the null hypothesis $H_0 \sim N(25, 0.625)$, and using the test values listed below, use both the critical value method and the p-value method to perform the following hypothesis test at the 5% significance level:

H_0: $\mu = 25$
H_1: $\mu < 25$

Use the following values of $\overline{X}$ to calculate the test statistic:
(a) 22.78   (b) 24.32   (c) 20.01   (d) 21.77   (e) 23.26   (f) 24.03

Right-Tailed Hypothesis Test

Consider yet again our present example. It is known from past records that the mean amount of money spent buying lunch by the students in this class is $25. Suppose that our sample of 40 students yielded a mean of $27. Can we then conclude based on this sample that the mean amount of money spent on lunch has actually increased? We perform a right-tailed hypothesis test at the 5% level of significance. The hypotheses will be formulated as follows:

H_0: $\mu = 25$
H_1: $\mu > 25$

Performing a right-tailed hypothesis test at the 5% significance level is the same as testing the value under consideration to see if it lies within the 95% lower confidence interval. Recall from chapter 6 that the 95% lower confidence interval where the sample mean was 25, the sample size was 40, and the population standard deviation was known to be 5 was: $\mu \leq 25.305$.

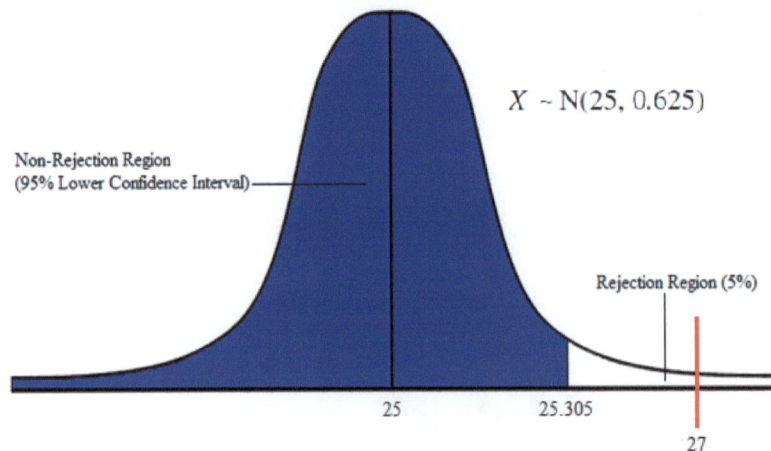

$X \sim N(25, 0.625)$

Non-Rejection Region
(95% Lower Confidence Interval)

Rejection Region (5%)

25   25.305

27

It is plain to see that the value 27 falls outside the 95% lower confidence interval, so we can therefore regard the difference between the sample mean of 27 and the original population mean 25 as a 'significant difference'. We therefore reject the null hypothesis and accept the alternative hypothesis, concluding that the mean has in fact increased.

As in the case of the two-tailed hypothesis test, we do not need the actual confidence interval in order to perform the hypothesis test. We can derive the Sampling Distribution of the Null Hypothesis, and then find the appropriate test statistic, which we will then place on the standard normal curve and use its location there to guide us to a solution.

The mean of the Sampling Distribution of the Null Hypothesis will be 25, which is the value of the null hypothesis (H_0: $\mu = 25$). The variance of this distribution will be equal to the known population variance (5^2), divided by the sample size (40).

So, $H_0 \sim N(25, \dfrac{5^2}{40}) \Rightarrow H_0 \sim N(25, 0.625)$

We then use this distribution to find the standardized z-value that would correspond to the test value of 27: This would be: $z_{27} = \dfrac{27 - 25}{\sqrt{0.625}} = \dfrac{27 - 25}{0.7906} = 2.53$.

The z-score on the standard normal curve that would give an area of 0.05 in the right tail is 1.64. Graphically, we now have:

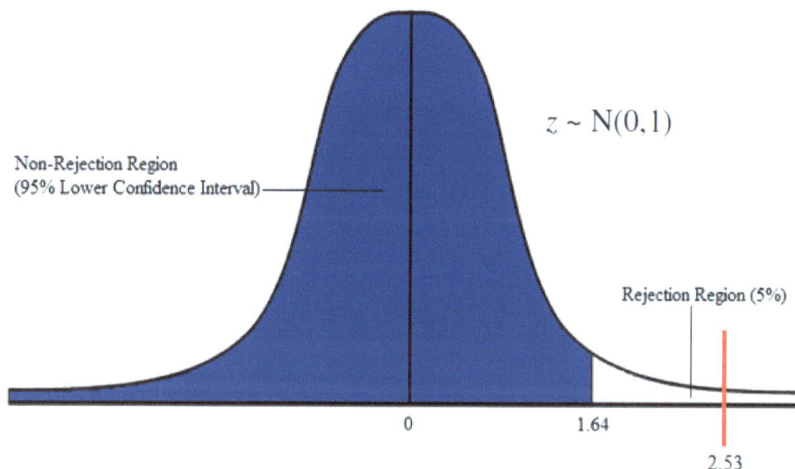

It is clear that the test statistic falls outside of the non-rejection region shaded in blue (the 95% lower confidence interval). Therefore we reject the null hypothesis and conclude that the new population mean as estimated by our sample mean of 27 is in fact greater than the original population mean of 25.

As an additional exercise, do this example using the p-value method. Verify that $p = 0.0057$, which is clearly less than the 'α' value of 0.05. We therefore reject the null hypothesis as above.

Exercise 7.4

For the sampling distribution of the null hypothesis $H_0 \sim N(25, 0.625)$, and using the values listed below, use both the critical value method and the p-value method to perform the following hypothesis test at the 5% significance level:

$H_0: \mu = 25$
$H_1: \mu > 25$

Use the following values of $\overline{X}$ to calculate the test statistic:
(a) 27.78   (b) 31.32   (c) 26.01   (d) 26.30   (e) 28.54   (f) 25.67

Exercise 7.5

(1) The average wingspan of a certain species of bird native to the Amazon rainforest was estimated at 1.35m in 1967. In 2007 some South American scientists measured the wingspans of a sample of 97 birds and found that the average wingspan of this sample was 1.28m, with a standard deviation of 0.39m. Perform a hypothesis test at the 3% significance level to determine if the mean wingspan of the 2007 sample points to a decrease in the wingspan of the birds between 1967 and 2007.

(2) The average life expectancy of the dogs in a certain city was measured at 12.9 years in 1956. Fifty years later, the average lifespan of a sample of 35 dogs was determined to be 15.3 years, with a standard deviation of 4.93 years. Is this sample sufficient cause to form the opinion that the average life expectancy of the dogs of this city has increased over the fifty year period? Test at the 2% significance level.

(3) In 1980, during the early years of the evolution of soca music in Trinidad and Tobago, the average length of a soca song was 5.5 minutes. The lengths of 56 randomly selected soca songs were measured in 2008 and the average length of these songs was 4.8 minutes, with a standard deviation of 0.95 minutes. Does this sample offer enough statistical evidence of a decrease in the length of a soca song between 1980 and 2008? Test at the 1% significance level.

(4) The average weight of the chickens at a farm in Mausica is 5.7 pounds. Farmer Brown has decided to purchase a new brand of feed for use at the farm – the aim being to increase the weights of the chickens. Nine months after the introduction of this new feed, 75 chickens were randomly chosen and weighed. The mean weight of these 75 chickens was 6.2 pounds, with a standard deviation of 3.5 pounds. Will a significance test at the 9% level of significance justify Farmer Brown's suspicion that the new feed is effective?

(5) The average weekly contribution per member of a particular English parish of the Anglican church is £45.70. A priest at this parish suspects that the weekly contributions of his members are declining. In order to possibly confirm his intuitive diagnosis, he statistically analyses the contents of 44 random envelopes one day after mass. He finds that the average contribution of these 44 envelopes is £41.50, with a standard deviation of £16.63. At the 4% significance level, can the priest say with statistical conviction that the parish members' average individual contributions have in fact declined?

(6) In the years prior to the proliferation of online media, the average time that students at the University of Reading spent each week reading newspapers was 4 hours. One year long after online media had become a popular alternative to the traditional media, a sample of 81 students at the university spent an average of 2.5 hours per week reading news. The standard deviation of their reading times was 1.5 hours. Using a hypothesis test at the 3% significance level, determine whether or not we can say that students at the University of Reading actually spent less time reading news at the time that the sample was taken.

(7) The Petroleum Company of Trinidad and Tobago employs the services of independent contractors. Five years ago, the average value of a contract was $1,001,096. During an auditing exercise, the auditors pulled 65 random contracts for inspection. The mean value of these contracts was calculated to be $1,684,263. The standard deviation was $1,951,236. Perform a hypothesis test at the 1% significance level and use this as a basis for deciding whether or not there has been an increase in the value of these contracts.

(8) Ten years ago, the average age of first-time offenders at a juvenile detention center was calculated to be 15.2 years. A series of government social programs were launched soon after those findings. Ten years later, a government spokesperson boasts about the efficacy of those social programs, claiming that the average age of the first-time juvenile offenders had gone up. If a random sample of 53 first-time offenders have an average age of 16.3 years, and the standard deviation of these ages is 2.3 years, at the 2% significance level, would you say that the spokesperson is correct?

(9) At national swimming trials, the average time for the men's 100 meter freestyle was 53.32 seconds before the introduction of new technologically advanced swimsuits. The year after the introduction of the swimsuit, the times of 50 of the swimmers were selected at random by a sports commentator and their average times measured for quick comparison with the times of the previous year. The average times of these swimmers was found to be 52.93 seconds, with a standard deviation of 1.04 seconds. At the 1% significance level, does this offer sufficient statistical evidence that the new swimsuit assisted in the reduction of the 100 meter freestyle times?

(10) A Caribbean cable channel has determined that of the ninety minutes of their nightly news cast, viewers typically stay tuned in for an average of about 45 minutes. In effort to keep viewers tuned in for a longer time than this, the channel implements a host of new measures designed to make the nightly newscast more attractive to their viewers. In order to determine how effective these measures were, the channel conducts a survey of 113 random households across the Caribbean. From these sampled households, the mean time spent watching the nightly newscast was 55 minutes, with a standard deviation of 59 minutes. Perform a hypothesis test at (a) 4% significance level, and (b) the 3% significance level, and use these tests to judge the success of the cable channel's initiatives.

Testing the Difference Between Two Population Means

Sometimes we may need to determine whether or not our samples offer sufficient statistical evidence that the means of the two populations from which the samples were drawn are different. If the two population means are different, then one will be greater than the other, or we can say alternatively that one will be less than the other.

If the difference between the two sample means is 'significant', then we infer that the two population means are 'different'. If the difference between the sample means is 'not significant', then the two population means are deemed to be 'not different' and for all practical intents and purposes, they will be considered 'the same'. The hypothesis test is also useful in this type of situation. For our purposes, we will consider only the cases of large samples drawn from populations that are independent of each other.

Example 7.1

A sample of 55 European films was found to have a mean length of 105 minutes with a standard deviation of 7 minutes. A sample of 62 Hollywood films was found to have a mean length of 90 minutes with a standard deviation of 5 minutes. Perform the appropriate hypothesis tests at the 5% significance level to test the following assumptions:

(a) European and Hollywood films are the same length
(b) European films are longer than Hollywood films
(c) Hollywood films are shorter than European films

Solution

(a) As always when performing hypothesis tests, the null hypothesis is the one that asserts 'no significant difference'. So in this case, our null hypothesis will suggest that there is 'no significant difference' between the length of a European film and the length of a Hollywood film - in other words, the length of a European film is equal to the length of a Hollywood film. Since we are simply interested in the existence of a difference without being concerned with the nature of the difference, the alternative hypothesis will simply assert that the length of a European film is not equal to the length of a Hollywood film.

For both samples, the distribution of the population is unknown, but since the sample size is greater than 30 in both cases, we have statistically large samples (Central Limit Theorem). The distributions of these sample means would therefore follow the normal distribution, with means equal to 105 and 90 minutes respectively for European and Hollywood films. The hypotheses are framed in terms of the entire set of European and Hollywood films (the populations) although in fact it will be the samples and their sample statistics that we will be using to carry out the test and draw the appropriate conclusions.

The null hypothesis can be expressed either in terms of the equality of the population means, or in terms of a difference between the population means. The two forms mean the same thing, but expressing the hypotheses in terms of a difference between population means makes it easier to follow the formation of the sampling distribution of the null hypothesis. If the two population means are equal, then the difference between them will be zero. If the two population means are not equal, then the difference between them will not be equal to zero.

The hypotheses are:

H$_0$: μ$_E$ = μ$_H$
H$_1$: μ$_E$ ≠ μ$_H$
    or:
H$_0$: μ$_E$ - μ$_H$ = 0   *(This form will facilitate the construction of the Sampling*
H$_1$: μ$_E$ - μ$_H$ ≠ 0   *Distribution of the Null Hypothesis)*

In both cases, the population variance is unknown, so we use the next available option, which is the variance of the respective samples. In each case, the sample variance is used as a point estimate of the population variance. We then obtain the variance of each sampling distribution by dividing the point estimate of the population variance by the sample size. So the respective variances would be $\frac{7^2}{55}$ for European films, and $\frac{5^2}{62}$ for Hollywood films. We therefore construct distributions for the samples of European and Hollywood films, and we will use these distributions to construct the Sampling Distribution of the Null Hypothesis. *As always*, we first define the random variables:

Let $\overline{E}$ be the randon variable 'mean length of a sample of 55 European films'

$$\overline{E} \sim N(105, \frac{7^2}{55})$$

Let $\overline{H}$ be the random variable 'mean length of a sample of 62 Hollywood films'

$$\overline{H} \sim N(90, \frac{5^2}{62})$$

We now proceed to construct the Sampling Distribution of the Null Hypothesis. The mean of this distribution is equal to the value of the null hypothesis, which is 0 (**H$_0$: μ$_E$ - μ$_H$ = 0**). The variance of this distribution is equal to the sum of the two variances in question.

$$\therefore \ E(H_0) = 0$$
$$\text{Var}(H_0) = \text{Var}(\overline{E}) + \text{Var}(\overline{H})$$
$$= \frac{7^2}{55} + \frac{5^2}{62}$$
$$= 0.8909 + 0.4032$$
$$= 1.2941$$

$$\therefore \ H_0 \sim N(0, 1.2941)$$

This is the implied sampling distribution from which a 95% Confidence Interval for the difference between the length of a European film and the length of a Hollywood film would have been constructed using our available sample statistics from a sample of 55 European films and a sample of 62 Hollywood films.

$$TV = E(\overline{E}) - E(\overline{H})$$
$$= 105 - 90$$
$$= 15$$

What we are in fact doing is testing this value to see whether or not it lies within the 95% Confidence Interval for the difference between the mean length of a European film and a Hollywood film. We do this by standardizing this test value on the Sampling Distribution of the Null Hypothesis and determining where it lies:

$$TS = \frac{15 - 0}{\sqrt{1.2941}} = \frac{15}{1.1376} = 13.19$$

So, our test statistic is 13.19, which we place on the standard normal distribution as:

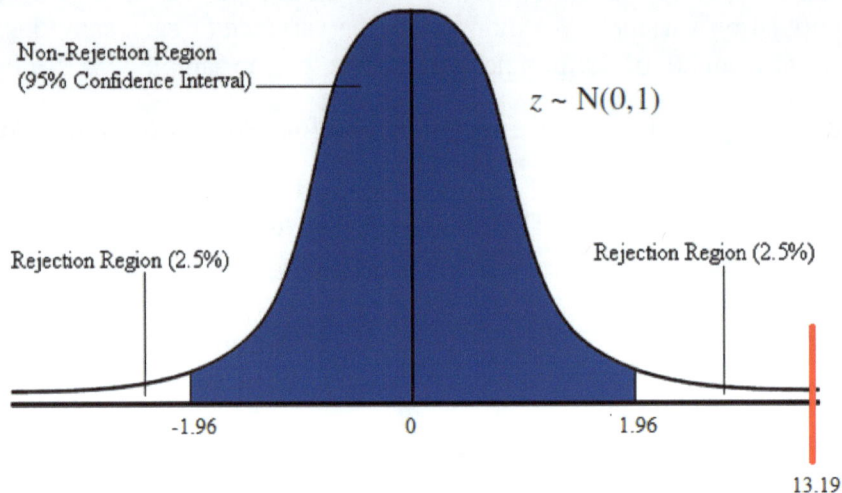

Our test statistic very clearly lies within the rejection region, so we reject the null hypothesis, in which case our conclusion would be that there is in fact a 'significant difference' in length between a European film and a Hollywood film. In other words, Hollywood and European films are not the same length.

We can use the p-value method here as well. For a standardized z-value of 13.19, $p \approx 0.0000$. This p-value is obviously less than the $\frac{\alpha}{2}$ value of 0.025, so we reject the null hypothesis, arriving at the same conclusions just spelt out.

(b) The null hypothesis can be expressed either in terms of the equality of the population means, or in terms of a difference between population means. The two forms mean the same thing, but expressing the hypotheses in terms of a difference between population means makes it easier to follow the formation of the sampling distribution of the null hypothesis. If the two population means are equal, then the difference between them will be zero. If the population mean for European films is greater than the population mean for Hollywood films, then when we subtract the Hollywood mean from the European mean, the difference will be greater than zero. The hypotheses in this case would be:

H_0: $\mu_E = \mu_H$
H_1: $\mu_E > \mu_H$
    or:
H_0: $\mu_E - \mu_H = 0$ *(This form will facilitate the construction of the Sampling*
H_1: $\mu_E - \mu_H > 0$ *Distribution of the Null Hypothesis)*

The Sampling Distribution of the Null Hypothesis would be the same in this case:
$\therefore \ H_0 \sim N(0, 1.2941)$

The test value in this instance will also be the same as before:
$$TV = E(\overline{E}) - E(\overline{H})$$
$$= 105 - 90$$
$$= 15$$

The Test Statistic too will also be the same:
$$TS = \frac{15 - 0}{\sqrt{1.2941}} = \frac{15}{1.1376} = 13.19$$

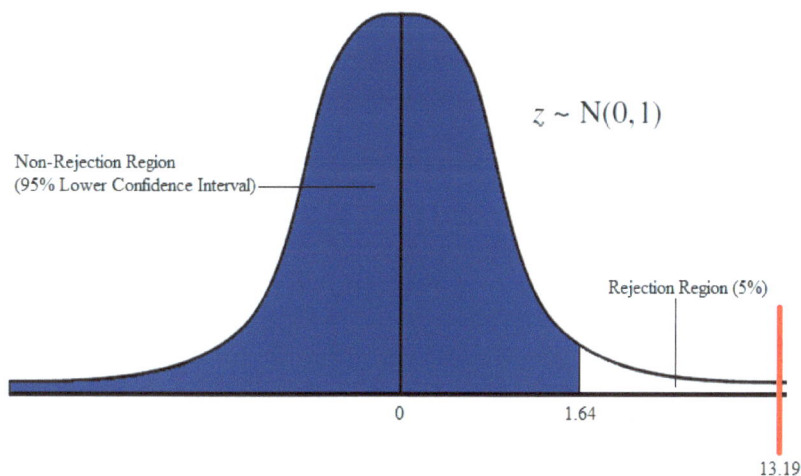

Our test statistic very clearly lies within the rejection region, so we reject the null hypothesis, in which case our conclusion would be that there is in fact a 'significant difference' in length between a Hollywood film and a European film, this difference being specifically one where the length of European films is greater than the length of Hollywood films. In other words, European films are longer than Hollywood films.

We can use the p-value method here as well. For a standardized z-value of 13.19, $p \approx 0.0000$. This p-value is obviously less than the α value of 0.05, so we reject the null hypothesis, arriving at the same conclusions just spelt out.

(c) The null hypothesis can be expressed either in terms of the equality of the population means, or in terms of a difference between population means. The two forms mean the same thing, but expressing the hypotheses in terms of a difference between population means makes it easier to follow the formation of the sampling distribution of the null hypothesis. If the two population means are equal, then the difference between them will be zero. If the population mean for Hollywood films is less than the population mean for European films, then when we subtract the European mean from the Hollywood mean, the difference will be less than zero.

The hypotheses in this case would be:

H_0: $\mu_H = \mu_E$

H_1: $\mu_H < \mu_E$

    or:

H_0: $\mu_H - \mu_E = 0$ *(This form will facilitate the construction of the Sampling*

H_1: $\mu_H - \mu_E < 0$ *Distribution of the Null Hypothesis)*

The Sampling Distribution of the Null Hypothesis would be the same in this case:

$\therefore$ $H_0 \sim N(0, 1.2941)$

The test value in this instance will also be the same as before:

$$TV = E(\overline{H}) - E(\overline{E})$$
$$= 90 - 105$$
$$= -15$$

The Test Statistic too will also be the same:

$$TS = \frac{-15-0}{\sqrt{1.2941}} = \frac{-15}{1.1376} = -13.19$$

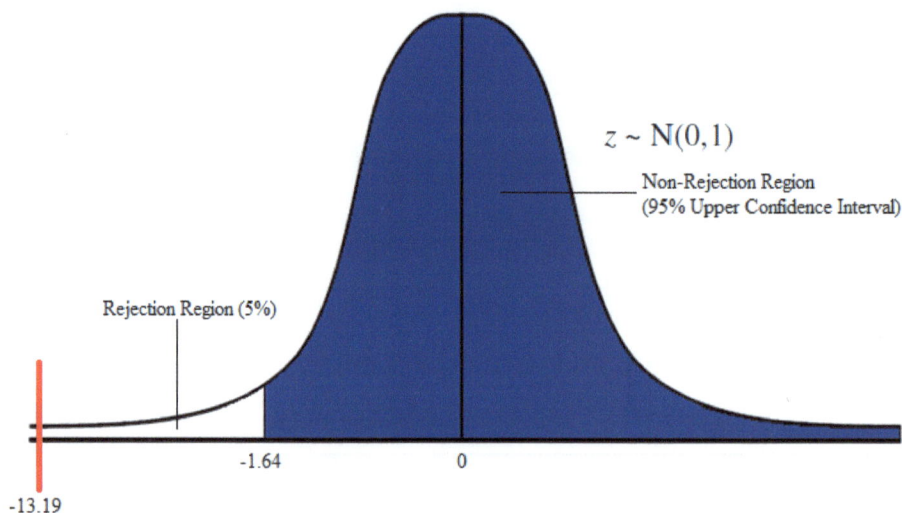

$z \sim N(0,1)$

Non-Rejection Region
(95% Upper Confidence Interval)

Rejection Region (5%)

-1.64   0

-13.19

Our test statistic very clearly lies within the rejection region, so we reject the null hypothesis, in which case our conclusion would be that there is in fact a 'significant difference' in length between a Hollywood film and a European film, this difference being specifically one where the length of Hollywood films is less than the length of European films. In other words, Hollywood films are shorter than European films.

We can use the p-value method here as well. For a standardized z-value of -13.19, $p \approx 0.0000$. This p-value is obviously less than the α value of 0.05, so we reject the null hypothesis, arriving at the same conclusions just spelt out.

Note that the alternative hypotheses in (**b**) and (**c**) are two diametrically opposed ways of saying the **SAME** thing! In other words, whether we perform a right-tailed or a left-tailed hypothesis test is dependent only on the specific linguistic and descriptive vagaries of each situation.

Exercise 7.6

(1) The average beats-per-minute of a sample of 45 soca songs from Trinidad and Tobago is 156, with a standard deviation of 26 beats per minute. The average beats-per-minute of a sample of 43 reggae songs from Jamaica is 112, with a standard deviation of 21 beats per minute. Test at the 2% significance level, the hypothesis that Jamaican reggae music is slower than soca music from Trinidad and Tobago.

(2) At the Scarborough hospital in Tobago, the mean weight of a sample of 87 newborn baby boys is 7.37 pounds, with a standard deviation of 0.87 pounds. The corresponding mean weight of a sample of 76 newborn baby girls is 7.19 pounds, with a standard deviation of 0.43 pounds. Perform an appropriate hypothesis test at the 4% significance level to examine the assumption that newborn boys at the Scarborough Hospital weigh more than newborn girls.

(3) The mean score by a basketball team in a 56 randomly chosen games played away from home is 76. The standard deviation of these scores is 6.4. The same team had a mean score of 83 with a standard deviation of 2.6 from a random sample of 37 games played at home. Test at the 1% level of significance the hypothesis that there is a significant difference between the average score in home games and the average score in away games for this team.

(4) The employees of Medford's Gas Station are having a friendly debate about the average sales for gasoline versus the average sales for diesel. They choose the next 70 gasoline sales, together with the next 60 diesel sales for comparison. The average gasoline sale was $66.36 with a standard deviation of $12.05. The average diesel sale was $71.83 with a standard deviation of $9.75. Will an appropriate hypothesis test at the 1% significance level support the claim by some employees that the drivers who purchase gasoline generally do so in smaller sales amounts?

(5) A counseling psychologist as part of his research for a new book on the differences between men and women is investigating the number of words spoken daily to each other by the men and women from a random selection of 100 married couples. The mean number of words spoken by the men was 2,000 with a standard deviation of 1500 words. The women on the other hand, spoke a mean of 4,500 words with a standard deviation of 850 words. Perform a hypothesis test at the 3% level of significance to determine if the psychologist's data supports the commonly held view that married women speak more than their husbands.

(6) As part of further research, the counseling psychologist investigates the behaviour of newly married couples and compares it with the behaviour of couples in their seventh year of marriage. He measures the frequency with which a sample of 48 newly married couples go out each month, and comes up with a mean frequency of 3.1 with a standard deviation of 0.5. Meanwhile, the mean frequency with which 48 couples in their seventh year of marriage go out is 1.9, with a standard deviation of 3.6. Perform a hypothesis test at the 2% level of significance to find out if these samples support the intuitional view that couples who have been married for seven years go out less frequently than couples who have recently been married.

(7) The mean time spent driving through traffic entering Port-of-Spain each morning by a sample of 38 drivers is 1.3 hours. The standard deviation of this time is 0.4 hours. The mean time spent driving through traffic leaving Port-of-Spain each evening by another sample of 42 drivers is 1.1 hours, with a standard deviation of 0.3 hours. Determine using a hypothesis test at (a) the 2% level of significance, and (b) the 1% level of significance, if there is any statistical difference between the time spent in traffic entering Port-of-Spain each morning and the time spent in traffic leaving Port-of-Spain each evening.

(8) The mean Mathematics score for a sample of 68 first year Economics students at the University of the West Indies is 58, with a standard deviation of 9.2. The mean Statistics score for another sample of 75 first year Economics students at the same university is 53, with a standard deviation of 4.6. (a) Will the appropriate hypothesis test at the 5% significance level allow us to conclude that the first year Economics students at the University of the West Indies performed worse in Statistics than they did in Mathematics? (b) Perform also an appropriate hypothesis test at the 5% level to test whether or not these samples indicate that these students performed better in Mathematics than they did in Statistics. Discuss the similarities and the differences between these two tests.

(9) As part of their investigations into recent observed changes in the yearly rainfall in Trinidad and Tobago, meteorologists are comparing daily rainfall in the first half of last year with the daily rainfall in the first half of this year. 60 days are chosen at random from the first half of each year and the rainfall measurements for each period analyzed. The mean and the standard deviation for the first half of last year are 2.3mm and 1.2mm of rainfall respectively. For the first half of this year, the mean and standard deviation are 2.9mm and 1.9mm of rainfall respectively. Conduct a hypothesis test at the 3% level of significance to determine if there is a 'significant' difference between the rainfall for the first half of the last year and the rainfall for the first half of this year.

(10) A political enthusiast in the United Kingdom believes that the members of the two main opposing parties speak for roughly the same amount of time. To test his belief, he measures the times of 62 speeches by members of the Conservative Party. These speeches have a mean of 43 minutes with a standard deviation of 12 minutes. A similar experiment carried out with 69 speeches by members of the Labour Party yields a mean of 49 minutes with a standard deviation of 9 minutes. Determine the conclusions to be drawn from a hypothesis test at the 1% significance level to investigate the difference between the speaking times of members of the Conservative Party and members of the Labour Party.

TESTING THE POPULATION PROPORTION
Two-Tailed Hypothesis Tests

We may know from past data that the proportion of students from the class of 100 who buy food at Top Lunch Café is 0.75. We take a sample of 40 students from our present class of 100, and find that 25 of them buy food at Top Lunch Café. The proportion of students from this sample who buy food at Top Lunch Café is therefore $\frac{25}{40}$ = 0.625. This is the sample proportion. We now perform a hypothesis test to investigate whether or not the population proportion has in fact changed.

Suppose we wish to test whether our sample proportion of 0.625 is an indication that the population proportion is now different from 0.75. Recall in chapter 6 that we had calculated a 95% confidence interval for the population proportion using a sample of size 40, a sample proportion of 0.75, and a sample variance of $\frac{(0.75)(0.25)}{40}$. If our sample proportion of 0.625 lies within this confidence interval, then we say that it is not significantly different from 0.75. We conclude then that our sample proportion of 0.625 does not indicate any change in the population proportion from the original value of 0.75. If our value of 0.625 falls outside of this confidence interval, then we say that it is significantly different from 0.75, and we therefore can say that the sample proportion of 0.625 indeed is an indication that there has been a change from the original population proportion of 0.75.

Because we are trying to see if the value 0.625 lies within a 95% confidence interval, that leaves us with a region of 5% outside of this interval. We are simply testing a difference without being specific as to the precise nature of that difference, so we must include in the test the case where the value being tested is less than the original value, as well as the case where the test value is greater than the original value. This means that a test value may either be 'significantly greater than' or 'significantly less than' the original value of 0.75. The 95% confidence interval would therefore be a two-sided confidence interval, implying a remaining region of 5% equally divided into two regions of 2.5% in either tail of the Normal Distribution Curve. Both regions of 2.5% together comprise the **region of significance**, giving us a hypothesis test at the 5% significance level (2 x 2.5%). The 95% confidence interval is therefore the **region of non-significance**.

The region of non-significance is also called the **non-rejection region**, since a test value and its associated test statistic falling in this region implies a difference that is 'not significant', and hence in that case we *do not reject* the null hypothesis. The region of significance is also called the **rejection region**, because once the test value and its associated test statistic falls within this region, the implied 'significant difference' means that we must *reject* the null hypothesis, accepting the alternative hypothesis as a result.

The null hypothesis (H_0) for this test is that there is no change in the previous population proportion (p = 0.75). The alternative hypothesis (H_1) would be that there has been a change in the population proportion (p $\neq$ 0.75). Recall from chapter 6, that our 95% confidence interval for the population proportion using a sample proportion of 0.75 and sample of size 40 was $0.6157 \leq p \leq 0.8843$. If the value of our sample proportion falls outside of this interval, then we can say that at the 5% significance level, our sample proportion is 'significantly different' from the original population proportion of 0.75.

If the value of the sample proportion falls within the interval, then we say that the sample proportion is 'not significantly different' from the original population proportion of 0.75. On this distribution, the sample proportion of 0.625 will fall inside the 95% confidence interval, as demonstrated in the graph.

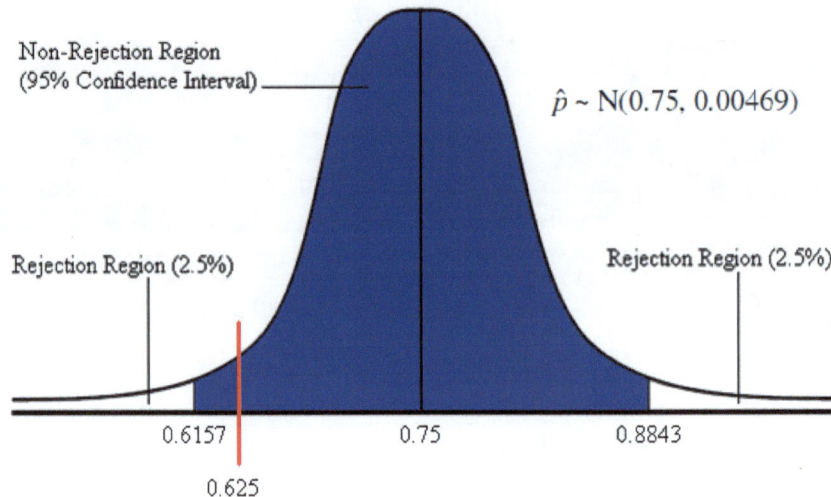

Non-Rejection Region
(95% Confidence Interval)

$\hat{p} \sim N(0.75, 0.00469)$

Rejection Region (2.5%)

Rejection Region (2.5%)

0.6157   0.75   0.8843

0.625

We therefore say that the value 0.625 is 'not significantly different' from the original population proportion of 0.75, in which case we say that the population proportion has NOT changed. In the jargon of hypothesis testing, we do not reject the null hypothesis.

As was the case with hypothesis tests for the population mean, when testing for the population proportion, we need not have prior knowledge of the corresponding confidence interval, neither do we need to know the relevant sampling distribution from which the corresponding confidence interval was derived. We can derive the real or implied sampling distribution, and we do not need to know the actual limits of the confidence interval that corresponds to the hypothesis test that we are doing. The first step would be to derive the real or implied distribution of the sample mean from which the confidence interval was derived:

(a) *Let the mean of this distribution be equal to the value of the null hypothesis*
(b) *Construct the variance of this distribution using the appropriate population parameters or sample statistics.*

This would be our **Sampling Distribution of the Null Hypothesis** as derived when we were testing the population mean.

In this case, following (a), the mean of the Sampling Distribution of the Null Hypothesis will be equal to 0.75, since this is the value of the null hypothesis $(\mathbf{H_0\!: p = 0.75})$. Following (b), we use the population proportion of 0.75 in tandem with the sample size of 40 to derive the variance as $\dfrac{(0.75)(0.25)}{40}$ in keeping with the theory on the variance of the distribution of the sample proportion.

The Sampling Distribution of the Null Hypothesis in this case therefore has a mean of 0.75, and a variance of $\frac{(0.75)(0.25)}{40}$. So, $H_0 \sim N(0.75, \frac{(0.75)(0.25)}{40}) \Rightarrow H_0 \sim N(0.75, 0.00469)$

Using this distribution, we now standardize the value of our sample proportion (0.625) to find the **Test Statistic**, which we will then place on the Standard Normal Distribution to see where it lies in relation to the **Critical Values**, which are the z-values on the Standard Normal Distribution that would give us the required areas in the region of significance or the rejection region.

A two-sided hypothesis test at the 5% significance level means that we would have a rejection region of 2.5% in each of the extremities of the Standard Normal Distribution curve. The z-values that will give us an area of 0.025 in each extremity would be -1.96 and 1.96. These are the critical values.

The test statistic is calculated as follows: $z_{0.625} = \dfrac{0.625 - 0.75}{\sqrt{0.00469}} = \dfrac{0.625 - 0.75}{0.0685} = -1.82$

On a graph we would have the following situation:

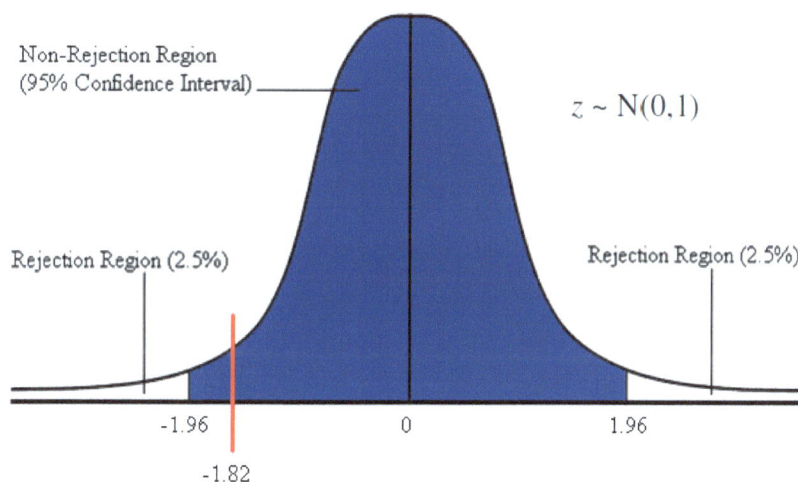

Clearly, our test statistic falls in the non-rejection region. We therefore DO NOT reject the null hypothesis, concluding on the basis of this test, that the original population proportion of 0.75 has remained unaltered. In other words, based on the results of the hypothesis test using our available sample, we conclude that the proportion of the class who buy at Top Lunch Café as estimated by our sample of size 40 is 'not significantly different' from 0.75 - in other words, it has not changed from 0.75.

Use of the p-value method

How do we apply the p-value method to this particular problem? Recall from page 113 our steps in applying the p-value method. Applying those steps here:

The critical values are -1.96 and +1.96. The test statistic is -1.82. We therefore compare the p-value of -1.82 to the $\frac{\alpha}{2}$ value of 0.025. The p-value for -1.82 is 0.0344. Graphically, we have:

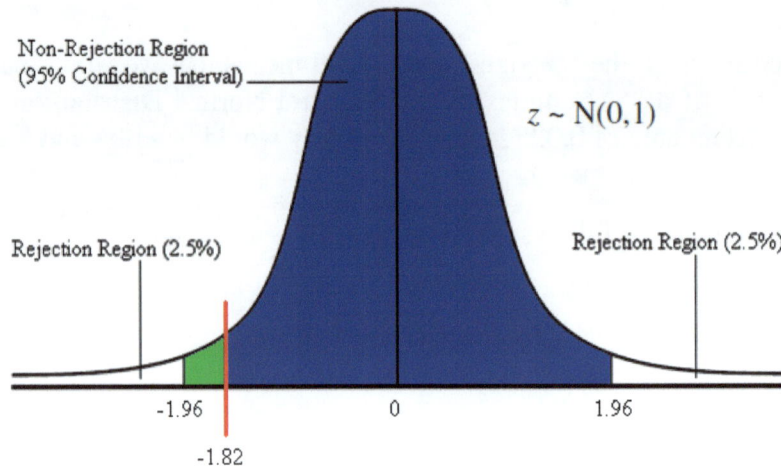

Non-Rejection Region
(95% Confidence Interval)

$z \sim N(0,1)$

Rejection Region (2.5%)

Rejection Region (2.5%)

-1.96   0   1.96

-1.82

The $\frac{\alpha}{2}$ value for -1.96 is the area in white which comprises the rejection region, while the p-value for -1.82 comprises the region in white (the $\frac{\alpha}{2}$ value) *in addition to* the region in green between -1.96 and -1.82. So clearly, the p-value for -1.82 is greater than the $\frac{\alpha}{2}$ value for -1.96.

This can only happen if the test statistic falls within the non-rejection region. So, since $p > \frac{\alpha}{2}$, we do not reject H_0. Our conclusion therefore is that the difference between the proportion of students from our sample who eat at Top Lunch Café and the proportion of students from the population who were known to eat at Top Lunch Café (past data) is not sufficiently significant to permit us to claim a change in the proportion of students who eat at Top Lunch Café.

In practice, graphs are not absolutely necessary when comparing the p-value with the $\frac{\alpha}{2}$ value (two-tailed test), or the α value (one-tailed test). We can simply compare them and make conclusions in line with the earlier note on page 114. However, as was done in these examples, *the use of graphs when doing hypothesis testing problems is strongly recommended!!*

Exercise 7.7

For the sampling distribution of the null hypothesis $H_0 \sim N(0.75, 0.00469)$ and using the test values listed below, use both the critical value method and the p-value method to perform the following hypothesis test at the 5% significance level:

H_0: $p = 0.75$
H_1: $p \neq 0.75$

Use the following values of $\hat{p}$:

(a) 0.66   (b) 0.83   (c) 0.712   (d) 0.786   (e) 0.58   (f) 0.94

Exercise 7.8 *(Express the sample proportions to two decimal places)*

(1) It is widely believed that 20% of the population at the University of Guyana is left-handed. During an exam, 13 of 53 students in a room are observed writing with their left hands. At the 20% significance level, do the students in this room suggest that the widespread belief is in fact true?

(2) It is commonly believed that about 90 percent of the East Indians in Trinidad and Tobago have East Indian surnames. A social researcher decides to carry out hypothesis tests at (a) the 3% significance level, and (b) the 1% significance level to start developing a statistically informed opinion. He surveys a random sample of 199 East Indians from across Trinidad and Tobago, and finds that 170 of them have East Indian surnames. What will the results of the tests reveal?

(3) The use of the Priority Bus Route in Trinidad is restricted to buses, taxis, and cars whose owners have special permission. In a road traffic exercise, the police pull over 80 drivers, 13 of whom do not have special permission to use the Priority Bus Route. Will the appropriate hypothesis test at the 7% significance level lend statistical support to the claim by the Commissioner of Police that one-quarter of the car drivers who use the Priority Bus Route do so illegally?

(4) The percentage of households with internet access in a small town in Grenada is rumored to be about 80% or thereabout. A random sample of 35 households from this town contains 33 homes with internet access. Based on these figures, can the rumor be described as 'true' at the 4% level of significance?

(5) The managers of a popular website claim that 67% of the daily visits to their website are from unique visitors. A random sample of 99 'site visits' on the website on a randomly chosen day revealed that 73 of the visits were from unique visitors. If we were to use the results from this sample as the basis for making a conclusion about the claims of the website's management, what would that conclusion be? Utilize a hypothesis test at the 8% significance level.

(6) An advertising campaign by the Bestel cellular phone network claims that only 5% of their calls are dropped. A consumer watchdog group decides to investigate the truth of this claim by monitoring 150 random calls on this network. Of these calls, 13 are dropped. What would the conclusion of the watchdog group be if the appropriate hypothesis test was performed at the 7% significance level?

(7) A group of rabidly angry housewives co-sign a letter to the Minister of National Security alleging that approximately one-quarter of the sacks of flour at Nice Food Supermarket are weevil-invested. The Minister commissions a ministry technocrat to investigate this claim. At the first stage of the investigation, the technocrat and his team inspect 70 randomly selected sacks of flour, 15 of which are found to be weevil-infested. Will a hypothesis test at the 2% level of significance support the claim of the housewives?

(8) The Board of Directors at a Roman Catholic High School suspects that just about half of the catholic students at the school attend Sunday Mass at least twice per month. They decide to test this hypothesis at the 6% significance level. A survey of 72 randomly selected students at the school reveal that 29 of those students surveyed attend mass at least twice a month. Do these results offer sufficient statistical support for the Board's suspicions?

(9) In touting the benefits of GPS technology for cars, The Minister of Works and Transport claims that approximately 10% of the cars in Trinidad and Tobago are outfitted with GPS technology. A clerk at the Licensing Authority decides to conduct a hypothesis test at the 3% significance level using the next 90 cars coming in to be licensed. In his observations, 87 of the cars do not have GPS technology. Does the data from this test resonate statistically with the claims of the Minister?

(10) Caribbean Airlines claims that 90 percent of their flights depart on time. 9 of a random sample of 71 Caribbean Airlines flights depart late. Is there sufficient statistical evidence at the 11% significance level to discount the airline's claim?

One-Tailed Hypothesis Tests

Suppose we wanted to get more specific than simply testing for a 'difference', or a 'change'? Let's say for example that we want to test for a 'negative change', a 'negative difference', or a 'decrease'. Alternatively, we may want to test for a 'positive change', a 'positive difference, or an 'increase'.

In the case of the university class of 100 students where a particular proportion of the students buy from Top Lunch Café, the null hypothesis would still be: **H_0: p = 0.75.** However, the alternative hypothesis would depend on the exact type of difference or change that we are testing for. In testing for a negative difference, we use a ***left-tailed test***, and in testing for a positive difference, we employ the ***right-tailed test.***

Left-Tailed Hypothesis Test

Consider again our present example. It is known from past records that the proportion of students from our class who buy food at Top Lunch Café is 0.75. Suppose that of our sample of 40, 25 students buy food at Top Lunch Café, giving a sample proportion of 0.625 ($\frac{25}{40}$). Can we then conclude based on this sample that the proportion of students who buy food at Top Lunch Café has actually decreased? We perform a left-tailed hypothesis test at the 5% level of significance. The hypotheses will be formulated as follows:

H$_0$: p = 0.75
H$_1$: p < 0.75

Performing a left-tailed hypothesis test at the 5% significance level is the same as testing the value under consideration to see if it lies within the 95% upper confidence interval. Recall from chapter 6 that the 95% upper confidence interval using a sample proportion of 0.75 and a sample size of 40 was: *p ≥ 0.6377*.

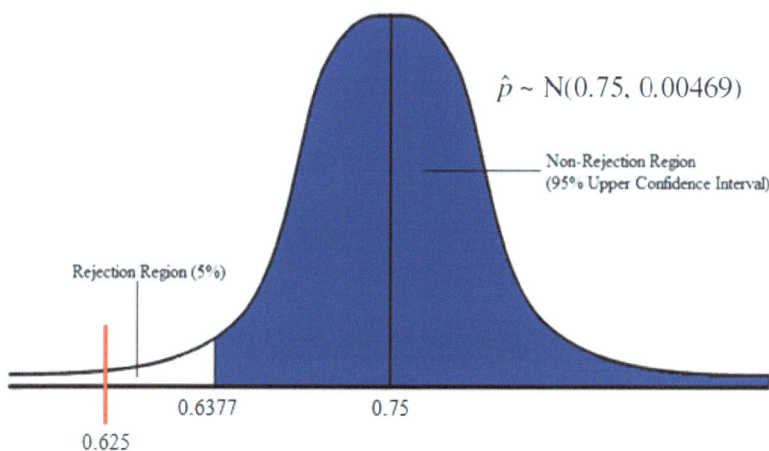

$\hat{p} \sim N(0.75, 0.00469)$

Non-Rejection Region
(95% Upper Confidence Interval)

Rejection Region (5%)

0.6377   0.75

0.625

Clearly, our sample proportion of 0.625 falls outside of the 95% upper confidence interval, so we can therefore regard the difference between the sample proportion of 0.625 and the original population proportion of 0.75 as a 'significant difference'. We therefore reject the null hypothesis and accept the alternative hypothesis, concluding that the proportion of students buying at Top Lunch Café has in fact decreased.

As in the case of the two-tailed hypothesis test, we do not need to have the actual confidence interval in order to perform the hypothesis test. We can derive the Sampling Distribution of the Null Hypothesis, and then find the appropriate test statistic, which we will then place on the standard normal curve and use its location there to guide us to a solution.

The mean of the Sampling Distribution of the Null Hypothesis in this instance will be 0.75, which is the value of the null hypothesis (**H$_0$: p = 0.75).** We derive the variance of the sampling distribution by using the population proportion of 0.75 in tandem with the sample size of 40 to derive the variance as $\frac{(0.75)(0.25)}{40}$ in keeping with the theory on the variance of the distribution of the sample proportion. The Sampling Distribution of the Null Hypothesis in this case will have a mean of 0.75, and a variance of $\frac{(0.75)(0.25)}{40}$.

So, $H_0 \sim N(0.75, \frac{(0.75)(0.25)}{40}) \Rightarrow H_0 \sim N(0.75, 0.00469)$

The test statistic is calculated as follows: $z_{0.625} = \frac{0.625 - 0.75}{\sqrt{0.00469}} = \frac{0.625 - 0.75}{0.0685} = -1.82$

The z-score on the standard normal curve that would give an area of 0.05 in the left tail is -1.64. Graphically, we now have:

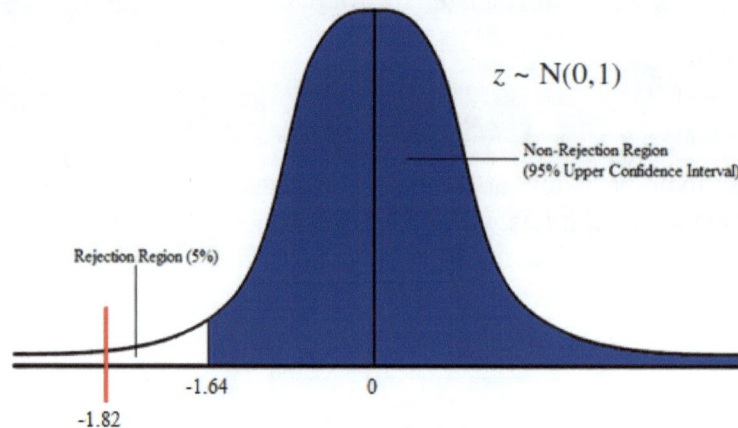

It is clear that the test statistic falls outside of the non-rejection region shaded in blue (the 95% upper confidence interval). It falls in the rejection region in the extremity of the left tail of the curve, and so we therefore reject the null hypothesis and conclude that the new population proportion as estimated by our sample proportion of 0.625 is in fact less than the original population proportion of mean of 0.75.

As an additional exercise, do this example using the p-value method. Verify that $p = 0.0057$, which is clearly less than the α value of 0.05. We therefore reject the null hypothesis as above.

Exercise 7.9

For the sampling distribution of the null hypothesis $H_0 \sim N(0.75, 0.00469)$, and using the test values listed below, use both the critical value method and the p-value method to perform the following hypothesis test at the 5% significance level:

H_0: p = 0.75
H_1: p < 0.75

Use the following values of $\hat{p}$ to calculate the test statistic:

(a) 0.69   (b) 0.73   (c) 0.712   (d) 0.597   (e) 0.65   (f) 0.638

Right-tailed Hypothesis Test

Consider again our present example. It is known from past records that the proportion of students from our class who buy food at Top Lunch Café is 0.75. Suppose that of our sample of 40, 32 students buy food at Top Lunch Café. This would be a sample proportion of $\frac{32}{40} = 0.8$.

Can we then conclude based on this sample that the proportion of students who buy at Top Lunch Café has now increased? We can perform a right-tailed hypothesis test at the 5% level of significance. The hypotheses will be formulated as follows:

H₀: p = 0.75
H₁: p > 0.75

Performing a left-tailed hypothesis test at the 5% significance level is the same as testing the value under consideration to see if it lies within the 95% lower confidence interval. Recall from chapter 6 that the 95% lower confidence interval was: *p ≤ 0.8623*.

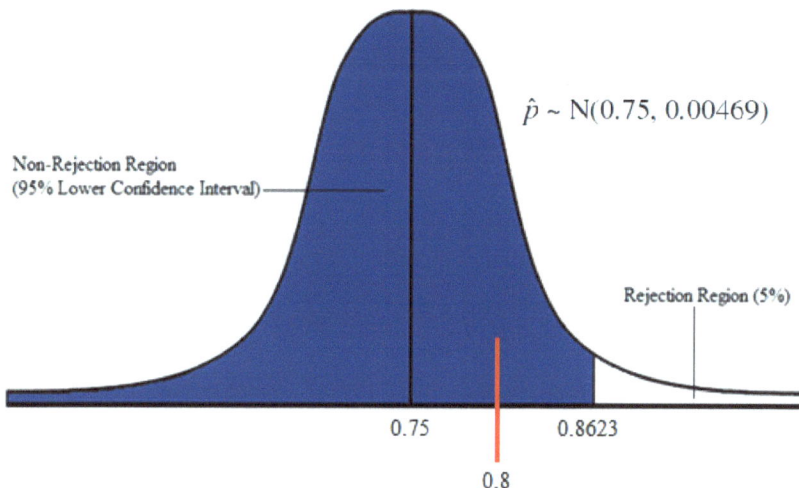

It is plain to see that the value 0.8 falls within the 95% lower confidence interval, so we can therefore regard the difference between the sample proportion of 0.8 and the original population proportion of 0.75 as a difference that is 'not significant'. We therefore do not reject the null hypothesis, concluding that the proportion of buyers at Top Lunch Café has remained unchanged.

As in the case of the two-tailed hypothesis test, we do not need to have the actual confidence interval in order to perform the hypothesis test. We can derive the Sampling Distribution of the Null Hypothesis, and then find the appropriate test statistic, which we will then place on the standard normal curve and use its location there to guide us to a solution.

The mean of the Sampling Distribution of the Null Hypothesis in this instance will be 0.75, which is the value of the null hypothesis (**H₀: p = 0.75**). We derive the variance of the sampling distribution by using the population proportion of 0.75 in tandem with the sample size of 40 to derive the variance as $\frac{(0.75)(0.25)}{40}$ in keeping with the theory on the variance of the distribution of the sample proportion. So the Sampling Distribution of the Null Hypothesis in this case has a mean of 0.75, and a variance of $\frac{(0.75)(0.25)}{40}$.

So, $H_0 \sim N(0.75, \frac{(0.75)(0.25)}{40}) \Rightarrow H_0 \sim N(0.75, 0.00469)$

The test statistic is calculated as follows: $z_{0.80} = \frac{0.80 - 0.75}{\sqrt{0.00469}} = \frac{0.80 - 0.75}{0.0685} = 0.73$

The z-score on the standard normal curve that would give an area of 0.05 in the right tail is 1.64. Graphically, we now have:

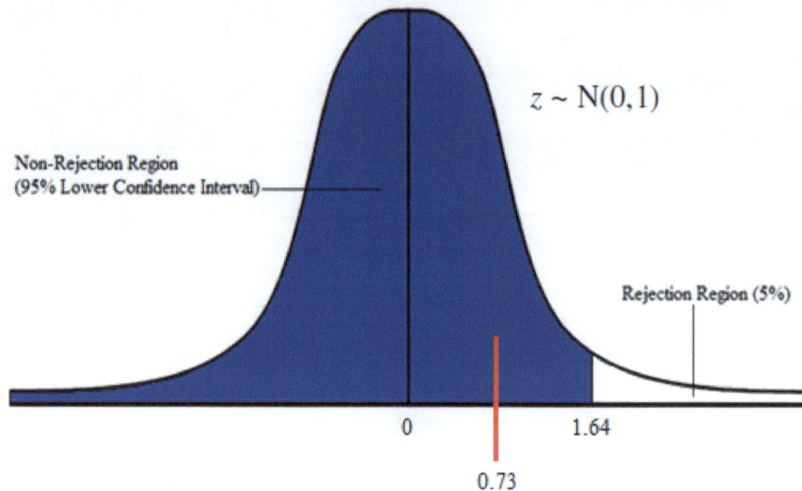

It is clear that the test statistic falls within the non-rejection region shaded in blue (the 95% lower confidence interval). Therefore we do not reject the null hypothesis and conclude that the new population proportion as estimated by our sample proportion of 0.80 is no different from the old population proportion of 0.75.

As an additional exercise, do this example using the p-value method. Verify that $p = 0.0057$, which is clearly less than the α value of 0.05. We therefore reject the null hypothesis as above.

Exercise 7.10

For the sampling distribution of the null hypothesis $H_0 \sim N(0.75, 0.00469)$, and using the test values listed below, use both the critical value method and the p-value method to perform the following hypothesis test at the 5% significance level:

H_0: p = 0.75
H_1: p > 0.75

Use the following values of $\hat{p}$ to calculate the test statistic:

(a) 0.79	(b) 0.777	(c) 0.812	(d) 0.90	(e) 0.83	(f) 0.864

Exercise 7.11 *(Express the sample proportions to two decimal places)*

(1) In 1968, one out of every four marriages in a certain country had ended in divorce within five years. In 2003, a group of psychologists started tracking 127 newlywed couples. By 2008, 41 of those couples were divorced. At the 4% level of significance, does this represent significant evidence of a general increase in the divorce rate in that country between 1968 and 2008?

(2) At a DVD factory in China, 6% of the DVDs manufactured contained defects. A quality assurance procedure was implemented, and a subsequent sample of 1,500 DVDs contained 60 defects. Is there sufficient justification at the 1% significant level for a claim that the quality assurance procedure was a success?

(3) In 1987, sixty percent of people getting married for the first time were under the age of 25. In 2007, a random sample of 200 people who married for the first time contained 96 people under the age of 25. At the 2% significance level, does this sample provide statistical confirmation of the popular perception that more people are getting married AFTER age 25?

(4) In 1956, 42% of the population of St.Vincent and the Grenadines was older than 65 years. In 2006, a random sample of the records of 499 members of the population contained 224 persons over the age of 65. If we perform the appropriate hypothesis test at the 8% significance level, does this sample safely allow us to conclude that there has been an increase in the proportion of the population over the age of 65 in that country?

(5) A particular course at the University of Technology in Jamaica has an average failure rate of 35%. In an effort to address this, the lecturer for the course decides to revise the course content and exam structure. In the first year after the revisions, a random sample of 135 students who wrote the exam contained 103 students who passed the course. At the 5% significance level, is this enough for the lecturer to claim success in his efforts to lower the failure rate?

(6) In 2006, police investigators determined that 57% of the traffic accidents in St.Lucia involved the use of alcohol by the driver. A vigorous "Don't Drink and Drive" public education campaign was launched in 2007. Half-way through 2008, a random sample of 100 road accidents included 46 that involved the use of alcohol by the driver. At (a) the 2% significance level, and (b) the 1% level of significance, will this data offer sufficient evidence that the 2007 "Don't Drink and Drive" campaign was effective?

(7) In the aftermath of a massive public outcry, the management of a radio station in Antigua are claiming to have increased the percentage of indigenous songs that are played on the airwaves. The logs for the last two years prior to the public outcry show that 19% of the songs played were indigenous songs. After the outcry, a random sample of 85 songs contained 23 indigenous songs. Is there enough statistical evidence at the 4% significance level to support the management's claim? Repeat the test at the 3% significance level.

(8) 15% of the e-mails coming to a particular e-mail account are 'spam' e-mails. After the installation of an anti-spam program designed to reduce the amount of spam e-mails, a random sample of 200 e-mails contains 24 that are categorized as spam. At the 9% significance level, does this data suggest that the anti-spam program is working?

(9) Farmer Franklin believes that about 10% of the cows on his farm are infected with mad cow disease. His wife Lady Franklin thinks that the infection rate is even higher. They both agree to test a sample of 57 cows in order to investigate the infection rate. 12 of the cows tested are infected with mad cow disease. If the test was done at the 1% significance level, who can we say is 'right'?

(10) A major commercial bank in the Bahamas reported an overall rate of loan default of 12 percent in the year before the world economic depression of 2009. In the immediate aftermath of this depression, a preliminary survey of 85 loans turned up 16 defaults. At the 2% significance level, does this suggest an increase in loan defaults since the 2009 economic depression?

Testing the Difference Between Two Population Proportions

Sometimes we may need to determine whether or not our samples offer sufficient statistical evidence that the proportions of two populations which possess a particular characteristic are the same, or whether they are different. If the two population proportions are found to be different, then one will be greater than the other, or we can say alternatively that one will be less than the other.

If the difference between the two sample proportions is 'significant', then we infer that the two population proportions are 'different'. If the difference between the sample proportions is 'not significant', then the two population proportions are deemed to be 'not different', and for all practical intents and purposes, they will be considered 'the same'. The hypothesis test is also useful in this type of situation. For our purposes, we will consider only the cases of large samples drawn from populations that are independent of each other.

Example 7.3

Of a random sample of 97 residents of Chaguanas, 36 were found to be smokers. A similar sample of 39 residents of the San Fernando contained 17 smokers. Perform the appropriate hypothesis tests at the 5% level to investigate the following assumptions:

(a) San Fernando and Chaguanas have the same proportion of smokers
(b) San Fernando has a greater proportion of smokers than Chaguanas
(c) Chaguanas has a lesser proportion of smokers than San Fernando

Solution

(a) As always when performing hypothesis tests, the null hypothesis is the one that asserts 'no significant difference'. So in this case, our null hypothesis will suggest that there is 'no significant difference' between the proportion of smokers in San Fernando and the proportion of smokers in Chaguanas, in other words, the proportion of smokers in San Fernando is equal to the proportion of smokers in Chaguanas. Since we are simply interested in the existence of a difference without being concerned with the nature of the difference, the alternative hypothesis will simply assert that the proportion of smokers in San Fernando is not equal to the proportion of smokers in Chaguanas.

The next step is to test the San Fernando and the Chaguanas samples to see whether or not they conform to the conditions that would allow the construction of the relevant distributions of the sample proportion. Recall from the definition of the Central Limit Theorem that these conditions are: $n\hat{p} > 5$, and $n\hat{q} > 5$.

For the San Fernando sample: $n = 39$; $\hat{p} = \dfrac{17}{39} = 0.44$; $\hat{q} = 1 - 0.44 = 0.56$

$$\therefore n\hat{p} = (39)(0.44) \qquad n\hat{q} = (39)(0.56)$$
$$= 17.16 \qquad\qquad = 21.84$$

And for the Chaguanas sample: $n = 97$; $\hat{p} = \dfrac{36}{97} = 0.37$; $\hat{q} = 1 - 0.37 = 0.63$

$$\therefore n\hat{p} = (97)(0.37) \qquad n\hat{q} = (97)(0.63)$$
$$= 35.89 \qquad\qquad = 61.11$$

In both cases, $n\hat{p}$ and $n\hat{q}$ are greater than 5, so we are free to construct the Distribution of the Sample Proportion for both the San Fernando and the Chaguanas samples.

As always, we first define the random variables.

Let $\hat{s}$ be the random variable 'proportion of a sample of 39 San Fernando residents who smoke'.
$\hat{p} = 0.44$, $\hat{q} = 0.56$, $n = 39$
$$\therefore \hat{s} \sim \mathrm{N}(0.44, \frac{(0.44)(0.56)}{39}) \Rightarrow \hat{s} \sim \mathrm{N}(0.44, 0.0063)$$

Let $\hat{c}$ be the random variable 'proportion of a sample of 97 Chaguanas residents who smoke'.
$\hat{p} = 0.37$, $\hat{q} = 0.63$, $n = 97$
$$\therefore \hat{c} \sim \mathrm{N}(0.37, \frac{(0.37)(0.63)}{97}) \Rightarrow \hat{c} \sim \mathrm{N}(0.37, 0.0024)$$

The hypotheses are framed in terms of the proportions of the entire populations of San Fernando and Chaguanas who smoke, although in fact it will be the samples from these locations that we will be using to test the equality of the proportions of smokers in the entire populations.

The null hypothesis can be expressed either in terms of the equality of the population proportions, or in terms of a difference between the population proportions. The two forms mean the same thing, but expressing the hypotheses in terms of a difference between population proportions makes it easier to follow the formation of the sampling distribution of the null hypothesis. If the two population proportions are equal, then the difference between them will be zero. If the two populations are not equal, then the difference between them will not be equal to zero. The hypotheses are:

H_0: $p_s = p_c$
H_1: $p_s \neq p_c$
   or:
H_0: $p_s - p_c = 0$ *(This form will facilitate the construction of the Sampling*
H_1: $p_s - p_c \neq 0$ *Distribution of the Null Hypothesis)*

We now proceed to construct the Sampling Distribution of the Null Hypothesis. The mean of this distribution is equal to the value of the null hypothesis, which is equal to 0. The variance of this distribution is equal to the sum of the two variances in question.

$$\therefore \ E(H_0) = 0$$

$$\text{Var}(H_0) = \text{Var}(\hat{s}) + \text{Var}(\hat{c})$$

$$= \frac{(0.44)(0.56)}{39} + \frac{(0.37)(0.63)}{97}$$

$$= 0.0063 + 0.0024$$

$$= 0.0087$$

$$\therefore \ H_0 \sim N(0, 0.0087)$$

The test value in this instance is the actual difference between the means of the distributions in question:

$$TV = E(\hat{s}) - E(\hat{c})$$

$$= 0.44 - 0.37$$

$$= 0.07$$

We now find the test statistic by standardizing the test value on the Sampling Distribution of the Null Hypothesis:

$$TS = \frac{0.07 - 0}{\sqrt{0.0087}} = \frac{0.07}{\sqrt{0.0087}} = \frac{0.07}{0.0933} = 0.75$$

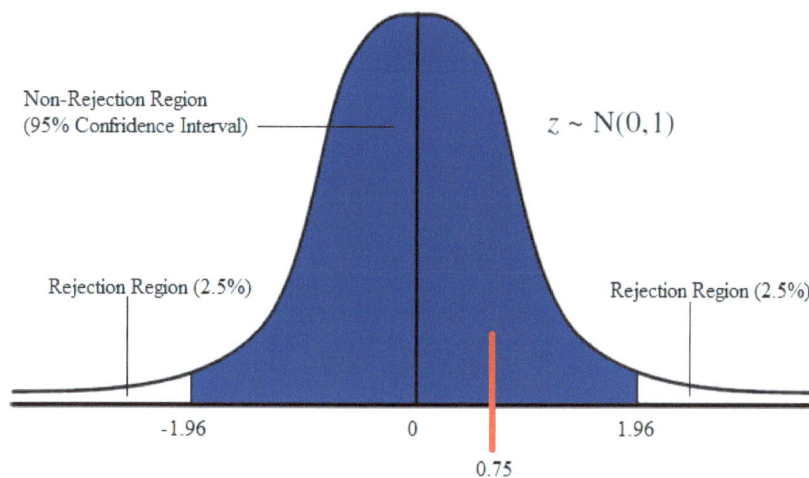

The test statistic lies within the non-rejection region. We therefore do not reject the null hypothesis, concluding that there is 'no significant difference' between the proportion of people in San Fernando who smoke and the proportion of people in Chaguanas who smoke. In other words, the proportion of smokers in San Fernando is the same as the proportion of smokers in Chaguanas.

Using the p-value method, for a standardized z-value of 0.75, $p = 0.2266$, which is greater than the $\dfrac{\alpha}{2}$ value of 0.025. We therefore do not reject the null hypothesis, arriving at the same conclusions just spelt out above.

(b) The null hypothesis can be expressed either in terms of the equality of the population proportions, or in terms of a difference between population proportions. The two forms mean the same thing, but expressing the hypotheses in terms of a difference between population means makes it easier to follow the formation of the sampling distribution of the null hypothesis. If the two population proportions are equal, then the difference between them will be zero. If the population proportion for San Fernando smokers is greater than the population proportion for Chaguanas smokers, then when we subtract the Chaguanas proportion from the San Fernando proportion, the difference will be greater than zero.

The hypotheses in this case would be:

H_0: $p_s = p_c$
H_1: $p_s > p_c$
   or:
H_0: $p_s - p_c = 0$ *(This form will facilitate the construction of the Sampling*
H_1: $p_s - p_c > 0$ *Distribution of the Null Hypothesis)*

The Sampling Distribution of the Null Hypothesis would be the same in this case:
∴ $H_0 \sim N(0, 0.0087)$

The test value will be the same as before:
$$TV = E(\hat{s}) - E(\hat{c})$$
$$= 0.44 - 0.37$$
$$= 0.07$$

The Test Statistic too will also be the same:

$$TS = \frac{0.07 - 0}{\sqrt{0.0087}} = \frac{0.07}{\sqrt{0.0087}} = \frac{0.07}{0.0933} = 0.75$$

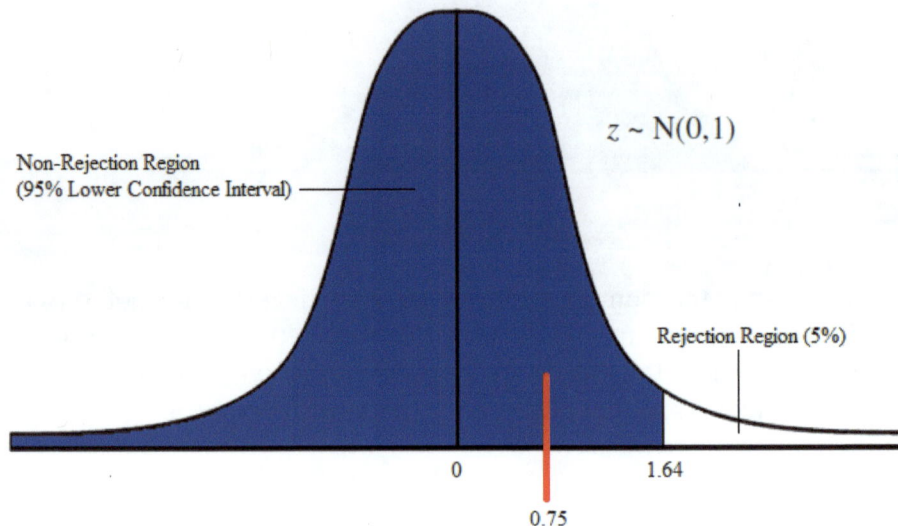

$z \sim N(0,1)$

Non-Rejection Region
(95% Lower Confidence Interval)

Rejection Region (5%)

0

1.64

0.75

The test statistic lies within the non-rejection region. We therefore do not reject the null hypothesis, concluding that there is 'no significant difference' between the proportion of people in San Fernando who smoke and the proportion of people in Chaguanas who smoke. In other words, the proportion of smokers in San Fernando is the same as the proportion of smokers in Chaguanas.

Using the p-value method, for a standardized z-value of 0.75, $p = 0.2266$, which is greater than the α value of 0.05. We therefore do not reject the null hypothesis, arriving at the same conclusions just spelt out above.

(c) The null hypothesis can be expressed either in terms of the equality of the population proportions, or in terms of a difference between population proportions. The two forms mean the same thing, but expressing the hypotheses in terms of a difference between population means makes it easier to follow the formation of the sampling distribution of the null hypothesis. If the two population proportions are equal, the difference between them will be zero. If the population proportion for Chaguanas smokers is less than the population proportion for San Fernando smokers, then when we subtract the San Fernando proportion from the Chaguanas proportion, the difference will be less than zero.

The hypotheses in this case would be:

H_0: $p_s = p_c$
H_1: $p_s < p_c$
    or:
H_0: $p_s - p_c = 0$ *(This form will facilitate the construction of the Sampling*
H_1: $p_s - p_c < 0$ *Distribution of the Null Hypothesis)*

The Sampling Distribution of the Null Hypothesis would be the same in this case:

$\therefore \ H_0 \sim N(0, 0.0087)$

The test value will be the same as before:

$TV = E(\hat{c}) - E(\hat{s})$

$\quad = 0.37 - 0.44$

$\quad = -0.07$

The Test Statistic too will also be the same as before:

$$TS = \frac{-0.07 - 0}{\sqrt{0.0087}} = \frac{-0.07}{\sqrt{0.0087}} = \frac{-0.07}{0.0933} = -0.75$$

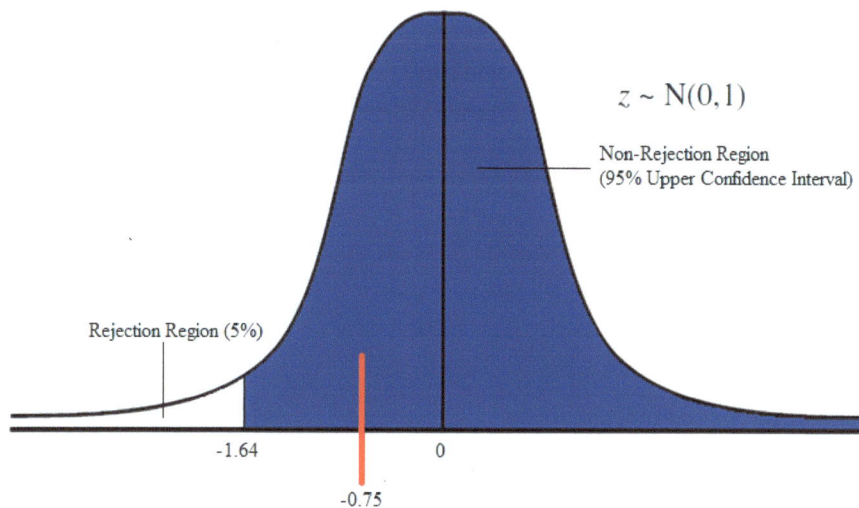

The test statistic lies within the non-rejection region. We therefore do not reject the null hypothesis, concluding that there is 'no significant difference' between the proportion of people in San Fernando who smoke and the proportion of people in Chaguanas who smoke. In other words, on the evidence of these two samples, the proportion of smokers in San Fernando is the same as the proportion of smokers in Chaguanas.

Using the p-value method, for a standardized z-value of -0.75, $p = 0.2266$, which is greater than the α value of 0.05. We therefore do not reject the null hypothesis, arriving at the same conclusions just spelt out above.

Note that the alternative hypotheses in (b) and (c) are two diametrically opposed ways of saying the **SAME** thing! In other words, whether we take a right-tailed or a left-tailed hypothesis test is dependent only on the specific linguistic vagaries of each situation.

Exercise 7.12 *(Express the sample proportions to two decimal places)*

(1) Of a sample of 97 residents of Port-of-Spain, 33 were found to be smokers. A similar sample of 39 residents of San Fernando contained 22 smokers. Is there sufficient statistical evidence from these two samples to claim that San Fernando contains a greater proportion of smokers than Port-of-Spain? Test at the 1% significance level.

(2) In order to reduce the recidivism rate (the proportion of repeat offenders in the criminal justice system), the Commissioner of Prisons at the local prison implements a series of reforms geared towards the holistic rehabilitation of inmates. Before the programs were implemented, 21 out of a sample of 68 inmates entering the prison were repeat offenders. Three years later, a sample of 97 inmates entering the prison contained 29 repeat offenders. If we were to use these figures to make a judgment of the efficacy of the prison reforms, at the 3% significance level, can it be said that the prison reforms were successful?

(3) An employee of the Central Statistical Office decides to compare the ratio of men marrying for the first time after age 40 with the ratio of women marrying for the first time after age 40. From one sample of the records of 105 marriages, 18 marriages involved men who married for the first time after age 40. Another sample of 123 marriages turned up 15 marriages where the woman married for the first time after 40. Can the employee then conclude on the basis of these figures that there is a difference between the proportion of men who marry after 40 and the proportion of women who marry after age 40? Test at the 6% significance level.

(4) A leading manufacturer of computer chips in the United States has a manufacturing plant at home and another in China. A leading industry publication wishes to compare the proportion of defective chips manufactured in the USA with the proportion of defective chips manufactured in China. Of a sample of 135 chips from the US plant, 16 are found to be defective, while there are 9 defective chips out of a sample of 112 chips from the Chinese plant. Perform a hypothesis test at the 20% significance level to determine if there is any difference between the proportion of defective chips from the US plant and the proportion of defective chips from the Chinese plant.

(5) A new government coming to power in St.Lucia puts forward the position that their detection rate for petty crime is higher than the corresponding detection rate of the last government. 43 randomly selected cases selected under the new government reveals that 17 of them were detected, while 19 of 54 randomly selected cases under the previous government were detected. At the 9% significance level, is the claim of the new government accurate??

(6) At an orchid greenhouse in rural Thailand, a new brand of fertilizer is being introduced in an attempt to decrease the proportion of seedlings that do not grow to maturity. Before the fertilizer was first applied, 29 out of a sample of 95 monitored seedlings did not grow to maturity. After the first application of the fertilizer, a sample of 76 monitored seedlings yielded 13 seedlings that did not grow to maturity. At the 3% significance level, is there evidence that the fertilizer has been effective?

(7) A leading pharmaceutical company is conducting tests into the efficacy of its latest asthma medication, with the understanding that the medication will be released only if the proportion of test subjects whose symptoms are improved by the medication is 'significantly' greater than the proportion of test subjects whose symptoms are improved by the placebo. A sample of 135 test subjects who were treated with the medication contained 120 who responded positively. Another sample of 111 test subjects who were treated with the placebo contained 76 who responded positively. Based on the appropriate hypothesis test at the 7% level, should the medication be released?

(8) Half of a sample of 66 North Koreans believes that the two Koreas should pursue political re-unification. Three-fifths of a sample of 75 South Koreans believes that their country should pursue re-unification with its northern neighbor. Do these samples offer statistical evidence at the 2% significance level that less North Koreans than South Koreans want re-unification?

(9) In the Democratic primary of the 2008 USA Presidential Elections, 197 of a random sample of 235 voters in South Carolina supported Barack Obama. A similar sample of 198 voters in North Carolina contained 101 Obama supporters. Using this data, test at the 1%, 3%, 5%, and 7% levels the hypothesis that there was no difference in support for Barack Obama between the two Carolinas.

(10) The mobile phone company Bestel is conducting an analysis of its comparative popularity in each island of the twin-island republic of Trinidad and Tobago. 164 mobile phone users in Trinidad are sampled, and 100 of them use Bestel, while 66 of 99 mobile phone users in sampled in Tobago use Bestel. At the 5% level of significance, can we say that there is any difference between the proportion of Bestel customers in Trinidad and the proportion of Bestel customers in Tobago?

Appendix A

The Standard Normal Distribution Tables

z – table 1

The values in the body of the z-table give the probability that a randomly chosen data value lies between the mean and a value z standard deviations above the mean. They are also interpreted as the proportion of data values that lie between the mean and a value z standard deviations above the mean.

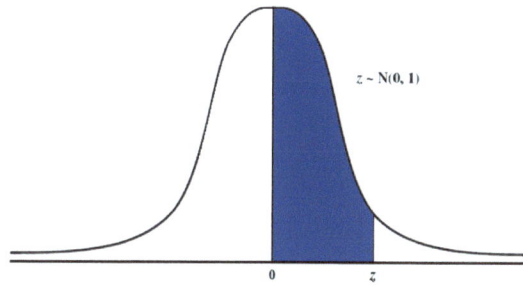

z ~ N(0, 1)

z	.00	.01	.02	.03	.04	.05	.06	.07	.08	.09
0.0	0.0000	0.0040	0.0080	0.0120	0.0160	0.0199	0.0239	0.0279	0.0319	0.0359
0.1	0.0398	0.0438	0.0478	0.0517	0.0557	0.0596	0.0636	0.0675	0.0714	0.0753
0.2	0.0793	0.0832	0.0871	0.0910	0.0948	0.0987	0.1026	0.1064	0.1103	0.1141
0.3	0.1179	0.1217	0.1255	0.1293	0.1331	0.1368	0.1406	0.1443	0.1480	0.1517
0.4	0.1554	0.1591	0.1628	0.1664	0.1700	0.1736	0.1772	0.1808	0.1844	0.1879
0.5	0.1915	0.1950	0.1985	0.2019	0.2054	0.2088	0.2123	0.2157	0.2190	0.2224
0.6	0.2257	0.2291	0.2324	0.2357	0.2389	0.2422	0.2454	0.2486	0.2517	0.2549
0.7	0.2580	0.2611	0.2642	0.2673	0.2704	0.2734	0.2764	0.2794	0.2823	0.2852
0.8	0.2881	0.2910	0.2939	0.2967	0.2995	0.3023	0.3051	0.3078	0.3106	0.3133
0.9	0.3159	0.3186	0.3212	0.3238	0.3264	0.3289	0.3315	0.3340	0.3365	0.3389
1.0	0.3413	0.3438	0.3461	0.3485	0.3508	0.3531	0.3554	0.3577	0.3599	0.3621
1.1	0.3643	0.3665	0.3686	0.3708	0.3729	0.3749	0.3770	0.3790	0.3810	0.3830
1.2	0.3849	0.3869	0.3888	0.3907	0.3925	0.3944	0.3962	0.3980	0.3997	0.4015
1.3	0.4032	0.4049	0.4066	0.4082	0.4099	0.4115	0.4131	0.4147	0.4162	0.4177
1.4	0.4192	0.4207	0.4222	0.4236	0.4251	0.4265	0.4279	0.4292	0.4306	0.4319
1.5	0.4332	0.4345	0.4357	0.4370	0.4382	0.4394	0.4406	0.4418	0.4429	0.4441
1.6	0.4452	0.4463	0.4474	0.4484	0.4495	0.4505	0.4515	0.4525	0.4535	0.4545
1.7	0.4554	0.4564	0.4573	0.4582	0.4591	0.4599	0.4608	0.4616	0.4625	0.4633
1.8	0.4641	0.4649	0.4656	0.4664	0.4671	0.4678	0.4686	0.4693	0.4699	0.4706
1.9	0.4713	0.4719	0.4726	0.4732	0.4738	0.4744	0.4750	0.4756	0.4761	0.4767
2.0	0.4772	0.4778	0.4783	0.4788	0.4793	0.4798	0.4803	0.4808	0.4812	0.4817
2.1	0.4821	0.4826	0.4830	0.4934	0.4838	0.4842	0.4846	0.4850	0.4854	0.4857
2.2	0.4861	0.4864	0.4868	0.4871	0.4875	0.4878	0.4881	0.4884	0.4887	0.4890
2.3	0.4893	0.4896	0.4898	0.4901	0.4904	0.4906	0.4909	0.4911	0.4913	0.4916
2.4	0.4918	0.4920	0.4922	0.4925	0.4927	0.4929	0.4931	0.4932	0.4934	0.4936
2.5	0.4938	0.4940	0.4941	0.4943	0.4945	0.4946	0.4948	0.4949	0.4951	0.4952
2.6	0.4953	0.4955	0.4956	0.4957	0.4959	0.4960	0.4961	0.4962	0.4963	0.4964
2.7	0.4965	0.4966	0.4967	0.4968	0.4969	0.4970	0.4971	0.4972	0.4973	0.4974
2.8	0.4974	0.4975	0.4976	0.4977	0.4977	0.4978	0.4979	0.4979	0.4980	0.4981
2.9	0.4981	0.4982	0.4982	0.4983	0.4984	0.4984	0.4985	0.4985	0.4986	0.4986
3.0	0.4987	0.4987	0.4987	0.4988	0.4988	0.4989	0.4989	0.4989	0.4990	0.4990

By virtue of the symmetric property of the Normal Distribution, the values in the body of the z-table also represent the probability that a randomly chosen data value lies between the mean and a value z standard deviations below the mean. Likewise, they are alternatively interpreted as the proportion of data values that lie between the mean and a data value z standard deviations below the mean. The standardized z-scores for values below the mean of a Normal Distribution are negative.

z – table 2

The values in the body of the z-table represent the probability that a randomly chosen data value lies to the right of a data value z standard deviations above the mean. They are also interpreted as the proportion of data values that are greater than a data value z standard deviations above the mean.

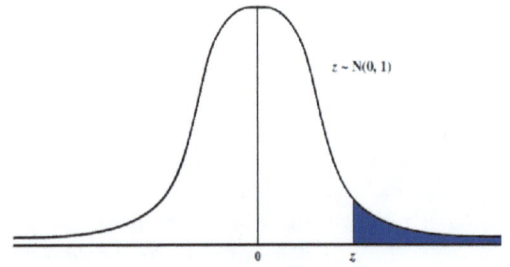

z	.00	.01	.02	.03	.04	.05	.06	.07	.08	.09
0.0	0.5000	0.4960	0.4920	0.4880	0.4840	0.4801	0.4761	0.4721	0.4681	0.4641
0.1	0.4602	0.4562	0.4522	0.4483	0.4443	0.4404	0.4364	0.4325	0.4286	0.4247
0.2	0.4207	0.4168	0.4129	0.4090	0.4052	0.4013	0.3974	0.3936	0.3897	0.3859
0.3	0.3821	0.3783	0.3745	0.3707	0.3669	0.3632	0.3594	0.3557	0.3520	0.3483
0.4	0.3446	0.3409	0.3372	0.3336	0.3300	0.3264	0.3228	0.3192	0.3156	0.3121
0.5	0.3085	0.3050	0.3015	0.2981	0.2946	0.2912	0.2877	0.2843	0.2810	0.2776
0.6	0.2743	0.2709	0.2676	0.2643	0.2611	0.2578	0.2546	0.2514	0.2483	0.2451
0.7	0.2420	0.2389	0.2358	0.2327	0.2296	0.2266	0.2236	0.2206	0.2177	0.2148
0.8	0.2119	0.2090	0.2061	0.2033	0.2005	0.1977	0.1949	0.1922	0.1894	0.1867
0.9	0.1841	0.1814	0.1788	0.1762	0.1736	0.1711	0.1685	0.1660	0.1635	0.1611
1.0	0.1587	0.1562	0.1539	0.1515	0.1492	0.1469	0.1446	0.1423	0.1401	0.1379
1.1	0.1357	0.1335	0.1314	0.1292	0.1271	0.1251	0.1230	0.1210	0.1190	0.1170
1.2	0.1151	0.1131	0.1112	0.1093	0.1075	0.1056	0.1038	0.1020	0.1003	0.0985
1.3	0.0968	0.0951	0.0934	0.0918	0.0901	0.0885	0.0869	0.0853	0.0838	0.0823
1.4	0.0808	0.0793	0.0778	0.0764	0.0749	0.0735	0.0721	0.0708	0.0694	0.0681
1.5	0.0668	0.0655	0.0643	0.0630	0.0618	0.0606	0.0594	0.0582	0.0571	0.0559
1.6	0.0548	0.0537	0.0526	0.0516	0.0505	0.0495	0.0485	0.0475	0.0465	0.0455
1.7	0.0446	0.0436	0.0427	0.0418	0.0409	0.0401	0.0392	0.0384	0.0375	0.0367
1.8	0.0359	0.0351	0.0344	0.0336	0.0329	0.0322	0.0314	0.0307	0.0301	0.0294
1.9	0.0287	0.0281	0.0274	0.0268	0.0262	0.0256	0.0250	0.0244	0.0239	0.0233
2.0	0.0228	0.0222	0.0217	0.0212	0.0207	0.0202	0.0197	0.0192	0.0187	0.0183
2.1	0.0179	0.0174	0.0170	0.0166	0.0162	0.0158	0.0154	0.0150	0.0146	0.0143
2.2	0.0139	0.0136	0.0132	0.0129	0.0126	0.0122	0.0119	0.0116	0.0113	0.0110
2.3	0.0107	0.0104	0.0102	0.0099	0.0096	0.0094	0.0091	0.0089	0.0087	0.0084
2.4	0.0082	0.0080	0.0078	0.0076	0.0073	0.0071	0.0070	0.0068	0.0066	0.0064
2.5	0.0062	0.0060	0.0059	0.0057	0.0055	0.0054	0.0052	0.0051	0.0049	0.0048
2.6	0.0047	0.0045	0.0044	0.0043	0.0041	0.0040	0.0039	0.0038	0.0037	0.0036
2.7	0.0035	0.0034	0.0033	0.0032	0.0031	0.0030	0.0029	0.0028	0.0027	0.0026
2.8	0.0026	0.0025	0.0024	0.0023	0.0023	0.0022	0.0021	0.0021	0.0020	0.0019
2.9	0.0019	0.0018	0.0018	0.0017	0.0016	0.0016	0.0015	0.0015	0.0014	0.0014
3.0	0.0013	0.0013	0.0013	0.0012	0.0012	0.0011	0.0011	0.0011	0.0010	0.0010

By virtue of the symmetric property of the Normal Distribution, the values in the body of the z-table also represent the probability that a randomly chosen data value lies to the left of a data value z standard deviations below the mean. Likewise, they are alternatively interpreted as the proportion of data values that are less than a data value z standard deviations below the mean. The standardized z-scores for values below the mean of a Normal Distribution are negative.

z – table 3

The values in the body of the z-table give the probability that a randomly chosen data value lies to the left of a data value z standard deviations above the mean. They are also interpreted as the proportion of data values that are less than a data value z standard deviations above the mean.

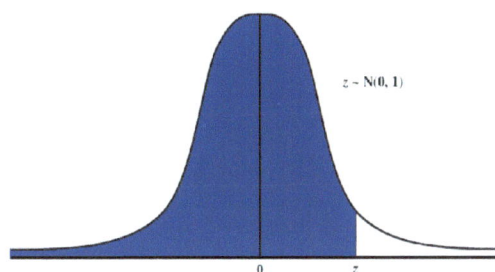

z ~ N(0, 1)

z	.00	.01	.02	.03	.04	.05	.06	.07	.08	.09
0.0	0.5000	0.5040	0.5080	0.5120	0.5160	0.5199	0.5239	0.5279	0.5319	0.5359
0.1	0.5398	0.5438	0.5478	0.5517	0.5557	0.5596	0.5636	0.5675	0.5714	0.5753
0.2	0.5793	0.5832	0.5871	0.5910	0.5948	0.5987	0.6026	0.6064	0.6103	0.6141
0.3	0.6179	0.6217	0.6255	0.6293	0.6331	0.6368	0.6406	0.6443	0.6480	0.6517
0.4	0.6554	0.6591	0.6628	0.6664	0.6700	0.6736	0.6772	0.6808	0.6844	0.6879
0.5	0.6915	0.6950	0.6985	0.7019	0.7054	0.7088	0.7123	0.7157	0.7190	0.7224
0.6	0.7257	0.7291	0.7324	0.7357	0.7389	0.7422	0.7454	0.7486	0.7517	0.7549
0.7	0.7580	0.7611	0.7642	0.7673	0.7704	0.7734	0.7764	0.7794	0.7823	0.7852
0.8	0.7881	0.7910	0.7939	0.7967	0.7995	0.8023	0.8051	0.8078	0.8106	0.8133
0.9	0.8159	0.8186	0.8212	0.8238	0.8264	0.8289	0.8315	0.8340	0.8365	0.8389
1.0	0.8413	0.8438	0.8461	0.8485	0.8508	0.8531	0.8554	0.8577	0.8599	0.8621
1.1	0.8643	0.8665	0.8686	0.8708	0.3729	0.8749	0.8770	0.8790	0.8810	0.8830
1.2	0.8849	0.8869	0.8888	0.8907	0.3925	0.8944	0.8962	0.8980	0.8997	0.9015
1.3	0.9032	0.9049	0.9066	0.9082	0.4099	0.9115	0.9131	0.9147	0.9162	0.9177
1.4	0.9192	0.9207	0.9222	0.9236	0.4251	0.9265	0.9279	0.9292	0.9306	0.9319
1.5	0.9332	0.9345	0.9357	0.9370	0.9382	0.9394	0.9406	0.9418	0.9429	0.9441
1.6	0.9452	0.9463	0.9474	0.9484	0.9495	0.9505	0.9515	0.9525	0.9535	0.9545
1.7	0.9554	0.9564	0.9573	0.9582	0.9591	0.9599	0.9608	0.9616	0.9625	0.9633
1.8	0.9641	0.9649	0.9656	0.9664	0.9671	0.9678	0.9686	0.9693	0.9699	0.9706
1.9	0.9713	0.9719	0.9726	0.9732	0.9738	0.9744	0.9750	0.9756	0.9761	0.9767
2.0	0.9772	0.9778	0.9783	0.9788	0.9793	0.9798	0.9803	0.9808	0.9812	0.9817
2.1	0.9821	0.9826	0.9830	0.9934	0.9838	0.9842	0.9846	0.9850	0.9854	0.9857
2.2	0.9861	0.9864	0.9868	0.9871	0.9875	0.9878	0.9881	0.9884	0.9887	0.9890
2.3	0.9893	0.9896	0.9898	0.9901	0.9904	0.9906	0.9909	0.9911	0.9913	0.9916
2.4	0.9918	0.9920	0.9922	0.9925	0.9927	0.9929	0.9931	0.9932	0.9934	0.9936
2.5	0.9938	0.9940	0.9941	0.9943	0.9945	0.9946	0.9948	0.9949	0.9951	0.9952
2.6	0.9953	0.9955	0.9956	0.9957	0.9959	0.9960	0.9961	0.9962	0.9963	0.9964
2.7	0.9965	0.9966	0.9967	0.9968	0.9969	0.9970	0.9971	0.9972	0.9973	0.9974
2.8	0.9974	0.9975	0.9976	0.9977	0.9977	0.9978	0.9979	0.9979	0.9980	0.9981
2.9	0.9981	0.9982	0.9982	0.9983	0.9984	0.9984	0.9985	0.9985	0.9986	0.9986
3.0	0.9987	0.9987	0.9987	0.9988	0.9988	0.9989	0.9989	0.9989	0.9990	0.9990

By virtue of the symmetric property of the Normal Distribution, the values in the body of the z-table also represent the probability that a randomly chosen data value lies to the right of a data value z standard deviations below the mean. Likewise, they are alternatively interpreted as the proportion of data values that are greater than a data value z standard deviations below the mean. The standardized z-scores for values below the mean of a Normal Distribution are negative.

Appendix B

Answers to Questions

Chapter 1

Exercise 1.1
(1) 5.5   **(2)** 9.72   **(3)** 0.0485   **(4)** -10   **(5)** 845.7   **(6)** 50.60   **(7)** -0.63
(8) 6,640.3   **(9)** -0.17   **(10)** 46.64

Exercise 1.2
(1) 2.06   **(2)** 5.38   **(3)** 0.0538   **(4)** 3.68   **(5)** 249.6   **(6)** 16.81
(7) 0.203   **(8)** 3,049.2   **(9)** 4.20   **(10)** 23.18

Exercise 1.3
(1) $X \sim N(23,14)$ or $X \sim N(23, 3.74^2)$
(2) $W \sim N(0.003, 0.12^2)$ or $W \sim N(0.003, 0.0144)$
(3) $P \sim N(1,256, 576)$ or $P \sim N(1,256, 24^2)$
(4) $Q \sim N(7,653,256, 1,245^2)$ or $Q \sim N(7,653,256, 1,572,516)$
(5) $Y \sim N(25.43, 16)$ or $Y \sim N(25.43, 4^2)$
(6) $B \sim N(-13.38, 3.68^2)$ or $B \sim N(-13.38, 13.54)$
(7) $F \sim N(2.337, 4.41)$ or $F \sim N(2.337, 2.1^2)$
(8) $K \sim N(24,643, 3,005^2)$ or $K \sim N(24,643, 9,030,025)$
(9) $T \sim N(-0.56, 0.99)$ or $T \sim N(-0.56, 0.99^2)$
(10) $V \sim N(-134, 12^2)$ or $V \sim N(-134, 144)$

Chapter 2

Exercise 2.1
(1) 0.25 **(2)** − 0.75 **(3)** 0 **(4)** 0.16 **(5)** − 0.47 **(6)** 1.25 **(7)** − 1.88 **(8)** − 2.25
(9) 2.25 **(10)** − 3.125

Exercise 2.2
(1) 0.54 standard deviations above the mean
(2) 1.21 standard deviations below the mean
(3) 0.99 standard deviations below the mean
(4) 1.92 standard deviations above the mean
(5) 2.42 standard deviations above the mean
(6) 3.01 standard deviations below the mean
(7) 2.31 standard deviations above the mean
(8) 0.56 standard deviations below the mean
(9) 1.13 standard deviations above the mean
(10) 2.07 standard deviations below the mean

Exercise 2.3
(1) 73.48 **(2)** 83.64 **(3)** 65 **(4)** 59.72 **(5)** 51.8 **(6)** 41.08 **(7)** 88.92 **(8)** 75.72
(9) 40.76 **(10)** 71.32

Exercise 2.4
(1) 0 standard deviations away from the mean
(2) 1.16 standard deviations above the mean
(3) 0.77 standard deviations below the mean
(4) 2.09 standard deviations above the mean
(5) 1.87 standard deviations below the mean
(6) 4.01 standard deviations above the mean
(7) 3.96 standard deviations below the mean
(8) 0.23 standard deviations above the mean
(9) 1.44 standard deviations below the mean
(10) 2.67 standard deviations below the mean

Exercise 2.5

(1) (a) 0.12   **(b)** 1.08   **(c)** - 0.36   **(d)** - 1.02

(2) (a) 1.52   **(b)** – 1.04   **(c)** 0.65   **(d)** – 0.74

(3) (a) 0.25   **(b)** – 0.5   **(c)** 0   **(d)** – 1.35   **(e)** 1.43

(4) (a) 1.74   **(b)** 0.56   **(c)** - 2.25   **(d)** 1.69   **(e)** – 2.38

(5) (a) -2.18   **(b)** 2.18   **(c)** 0.73   **(d)** 0   **(e)** - 1.27   **(f)** - 0.55
      (g) 1.82

(6) (a) 175.42   **(b)** 119.31   **(c)** 165.96   **(d)** 155   **(e)** 171.93

(7) (a) – 0.008   **(b)** 0.022   **(c)** 0.04   **(d)** 0.058   **(e)** 0.088   **(f)** 0.073

(8) (a) - 0.01   **(b)** – 0.085   **(c)** – 0.071   **(d)** 0.018   **(e)** – 0.049

(9) (a) 72   **(b)** 103.62   **(c)** 76.90   **(d)** 85.08

(10) (a) 189.53   **(b)** 191.50   **(c)** 166.05   **(d)** 173.35   **(e)** 190.51   **(f)** 164.73

Chapter 3

Exercise 3.1
(1) 0.1562   **(2)** 0.3050   **(3)** 0.0681   **(4)** 0.1977   **(5)** 0.0033   **(6)** 0.0012   **(7)** 0.0228
(8) 0.4721   **(9)** 0.0250   **(10)** 0.2546

Exercise 3.2
(1) 0.1562   **(2)** 0.3050   **(3)** 0.0934   **(4)** 0.4364   **(5)** 0.0080   **(6)** 0.0011   **(7)** 0.0015
(8) 0.0505   **(9)** 0.1611   **(10)** 0.0250

Exercise 3.3
(1) 0.3365   **(2)** 0.1556   **(3)** 0.0881   **(4)** 0.3360   **(5)** 0.0957   **(6)** 0.3503   **(7)** 0.0131
(8) 0.3571   **(9)** 0.1903   **(10)** 0.1359

Exercise 3.4
(1) 0.4089   **(2)** 0.1337   **(3)** 0.3845   **(4)** 0.4334   **(5)** 0.0617   **(6)** 0.3365   **(7)** 0.1556
(8) 0.3227   **(9)** 0.3021   **(10)** 0.1010

Exercise 3.5
(1) 0.7304   **(2)** 0.7802   **(3)** 0.5489   **(4)** 0.4492   **(5)** 0.8751   **(6)** 0.8968   **(7)** 0.5003
(8) 0.6986   **(9)** 0.9834   **(10)** 0.6866

Exercise 3.6
(1) 0.9778   **(2)** 0.6736   **(3)** 0.9783   **(4)** 0.8389   **(5)** 0.9265   **(6)** 0.9990   **(7)** 0.9964
(8) 0.5948   **(9)** 0.9633   **(10)** 0.8599

Exercise 3.7
(1) 0.9591   **(2)** 0.9750   **(3)** 0.7454   **(4)** 0.9938   **(5)** 0.8289   **(6)** 0.7324   **(7)** 0.9082
(8) 0.9960   **(9)** 0.9767   **(10)** 0.9987

Exercise 3.8
(1) 1.03　　**(2)** 0.84　　**(3)** 1.65　　**(4)** 0.58　　**(5)** 2.51　　**(6)** 1.82　　**(7)** 1.28
(8) 1.64 or 1.65　**(9)** 0.45　**(10)** 0.60

Exercise 3.9
(1) − 1.03　**(2)** − 0.79　**(3)** − 1.64　**(4)** − 2.44　**(5)** − 0.02　**(6)** − 0.96　**(7)** − 0.27
(8) − 3.03 or − 3.04　**(9)** − 0.22　**(10)** − 0.63

Exercise 3.10
(1) − 0.72　**(2)** − 0.25　**(3)** − 1.04　**(4)** − 0.13　**(5)** − 1.36　**(6)** − 0.35　**(7)** − 0.60
(8) − 0.94　**(9)** − 0.57　**(10)** − 1.43

Exercise 3.11
(1) 0.72　　**(2)** 0.25　　**(3)** 0.09　　**(4)** 1.01　　**(5)** 1.28　　**(6)** 0.35　　**(7)** 1.53
(8) 0.66　　**(9)** 0.24　　**(10)** 0.92

Exercise 3.12
(1) 0.2611　**(2)** 0.0885　**(3)** 0.0222　**(4)** 0.1446　**(5)** 0.1660　**(6)** 0.0735
(7) 0.1003　**(8)** 0.0051　**(9)** 0.0351　**(10)** 0.2611

Exercise 3.13
(1) -1.64 or -1.65　　**(2)** 1.32　　**(3)** 1.06　　**(4)** -0.25　　**(5)** -1.96　　**(6)** 2.33
(7) -2.57 or -2.58　　**(8)** -1.81　**(9)** 0.72　**(10)** 1.13

Exercise 3.14
(1) 0.2119　**(2)** 0.3469　　**(3)** 0.4090　**(4)** 80 or 81 (derived from a probability of 0.8031)
(5) 0.2389　**(6)** (i) 65.6 (ii) 49.2　**(7)** $2,878,241.88　**(8)** 201.44 pounds
(9) $302.44 to $653.24　**(10)** 44.06 beats per minute

Chapter 4

Exercise 4.1

(1) (a) $2X \sim N(210, 900)$ **(b)** $3X \sim N(315, 2025)$ **(c)** $2.3X \sim N(241.5, 1190.25)$
     (d) $4.06X \sim N(426.3, 3708.81)$

(2) (a) $0.5P \sim N(0.025, 0.0025)$ **(b)** $12P \sim N(0.6, 1.44)$ **(c)** $5.77P \sim N(0.2885, 0.332929)$
     (d) $7.5P \sim N(0.375, 0.5625)$

(3) (a) $14K \sim N(47.6, 237.16)$ **(b)** $6K \sim N(20.4, 43.56)$ **(c)** $8.21K \sim N(27.91, 81.56)$
     (d) $0.024K \sim N(0.0816, 0.00697)$

(4) (a) $2U \sim N(4698, 900)$ **(b)** $0.77U \sim N(1808.73, 133.40)$ **(c)** $0.04U \sim N(93.96, 0.36)$
     (d) $120U \sim N(281880, 3240000)$

(5) (a) $8F \sim N(432, 1296)$ **(b)** $12.47F \sim N(673.38, 3138.89)$ **(c)** $0.9F \sim N(48.6, 16.403)$
     (d) $265F \sim N(14310, 1422056.25)$

(6) (a) $X + Y \sim N(144, 52)$ **(b)** $X - Y \sim N(12, 52)$ **(c)** $2X + 2Y \sim N(288, 208)$
     (d) $2X - 2Y \sim N(24, 208)$

(7) (a) $P + Q \sim N(28, 13)$ **(b)** $P - Q \sim N(-4, 13)$ **(c)** $2P + Q \sim N(40, 25)$
     (d) $P - 2Q \sim N(-20, 40)$

(8) (a) $C - D \sim N(-0.23, 0.0269)$ **(b)** $D - C \sim N(0.23, 0.0269)$
     (c) $2D - 4C \sim N(-0.54, 0.2276)$ **(d)** $3C - 2D \sim N(0.04, 0.1576)$

(9) (a) $V - T \sim N(265, 7405)$ **(b)** $T - V \sim N(-265, 7405)$
     (c) $2T + V \sim N(4765, 16153)$ **(d)** $2T - V \sim N(1235, 16153)$

(10) (a) $G + H \sim N(758, 821)$ **(b)** $H + G \sim N(758, 821)$
     (c) $G - H \sim N(152, 821)$ **(d)** $H - G \sim N(-152, 821)$

Exercise 4.2

(1) (a) 0.3015 (b) 0.0465 (c) 1283.9   **(2)** 0.1179   **(3)** 0.8771   **(4)** 0.8665   **(5)** 0.0485
(6) 0.6406   **(7)** 0.0918   **(8)** 0.6554   **(9)** (i) 0.0081 (ii) 0.9946 (iii) 0.0516   **(10)** 0.1949

Chapter 5

Exercise 5.1

(1) (a) $\bar{X} \sim N(151, \dfrac{15^2}{45})$ or $\bar{X} \sim N(151, 5)$

  (b) $\bar{X} \sim N(151, \dfrac{15^2}{26})$ or $\bar{X} \sim N(151, 8.65)$

  (c) $\bar{X} \sim N(151, \dfrac{15^2}{64})$ or $\bar{X} \sim N(151, 3.52)$

  (d) (c) $\bar{X} \sim N(151, \dfrac{15^2}{17})$ or $\bar{X} \sim N(151, 13.26)$

(2) (a) $\bar{P} \sim N(0.3, \dfrac{0.2^2}{49})$ or $\bar{P} \sim N(0.3, 0.00082)$

  (b) $\bar{P} \sim N(0.3, \dfrac{0.2^2}{205})$ or $\bar{P} \sim N(0.3, 0.000195)$

  (c) $\bar{P} \sim N(0.3, \dfrac{0.2^2}{11})$ or $\bar{P} \sim N(0.3, 0.0036)$

  (d) $\bar{P} \sim N(0.3, \dfrac{0.2^2}{99})$ or $\bar{P} \sim N(0.3, 0.0040)$

(3) $\bar{X} \sim N(132.56, \dfrac{9.43^2}{63})$ or $\bar{X} \sim N(132.56, 1.41)$

(4) $\bar{X} \sim N(56.3, \dfrac{11}{22})$ or $\bar{X} \sim N(56.3, 0.5)$

(5) $\bar{X} \sim N(71.11, \dfrac{8.64^2}{96})$ or $\bar{X} \sim N(71.11, 6.77)$

(6) $\bar{X} \sim N(25.4, \dfrac{0.5^2}{40})$ or $\bar{X} \sim N(25.4, 0.00625)$

(7) $\bar{X} \sim N(80, \dfrac{2.67}{39})$ or $\bar{X} \sim N(80, 0.068)$

(8) $\bar{X} \sim N(5.66, \dfrac{1.8}{50})$ or $\bar{X} \sim N(5.66, 0.036)$

(9) $\bar{X} \sim N(0.12, \dfrac{0.02^2}{65})$ or $\bar{X} \sim N(0.12, 0.0000062)$

(10) $\bar{X} \sim N(0.063, \dfrac{0.014^2}{100})$ or $\bar{X} \sim N(0.063, 0.00000196)$

Exercise 5.2

(1) 2.28%, derived from a probability of 0.0228   **(2)** 0.4168   **(3)** 0.0485   **(4)** 0.0607
(5) approximately 0. This probability of '0' means that based on these statistics, none of the patients from a random sample of 40 can be expected to have a heart rate below 59 beats per minute.
(6) 0.2981   **(7)** 0.9910   **(8)** 0.1736   **(9)** 0.6846   **(10)** 0.2061

Exercise 5.3

(1) (a) $\hat{p} \sim N(0.5, \dfrac{(0.5)(0.5)}{30})$ or $\hat{p} \sim N(0.5, 0.00833)$

    (b) $\hat{p} \sim N(0.5, \dfrac{(0.5)(0.5)}{60})$ or $\hat{p} \sim N(0.5, 0.004167)$

    (c) $\hat{p} \sim N(0.5, \dfrac{(0.5)(0.5)}{99})$ or $\hat{p} \sim N(0.5, 0.2525)$

    (d) $\hat{p} \sim N(0.5, \dfrac{(0.5)(0.5)}{145})$ or $\hat{p} \sim N(0.5, 0043)$

(2) (a) $\hat{p} \sim N(0.49, \dfrac{(0.49)(0.51)}{61})$ or $\hat{p} \sim N(0.49, 0.004096)$

    (b) $\hat{p} \sim N(0.49, \dfrac{(0.49)(0.51)}{235})$ or $\hat{p} \sim N(0.49, 0.00106)$

    (c) $\hat{p} \sim N(0.49, \dfrac{(0.49)(0.51)}{115})$ or $\hat{p} \sim N(0.49, 0.00217)$

    (d) $\hat{p} \sim N(0.49, \dfrac{(0.49)(0.51)}{49})$ or $\hat{p} \sim N(0.49, 0.0051)$

(3) $\hat{p} \sim N(0.8, \dfrac{(0.8)(0.2)}{40})$ or $\hat{p} \sim N(0.8, 0040)$

(4) $\hat{p} \sim N(0.05, \dfrac{(0.05)(0.95)}{200})$ or $\hat{p} \sim N(0.05, 0.0002375)$

(5) $\hat{p} \sim N(0.6, \dfrac{(0.6)(0.4)}{39})$ or $\hat{p} \sim N(0.6, 0.00615)$

(6) (a) $\hat{p} \sim N(0.25, \dfrac{(0.25)(0.75)}{100})$ or $\hat{p} \sim N(0.25, 0.001875)$

    (b) $\hat{p} \sim N(0.63, \dfrac{(0.63)(0.37)}{100})$ or $\hat{p} \sim N(0.63, 0.002331)$

    (c) $\hat{p} \sim N(0.88, \dfrac{(0.88)(0.12)}{100})$ or $\hat{p} \sim N(0.88, 0.001056)$

(d) $\hat{p} \sim N(0.17, \frac{(0.17)(0.83)}{100})$ or $\hat{p} \sim N(0.17, 0.001411)$

(7) (a) $\hat{p} \sim N(0.50, \frac{(0.50)(0.50)}{126})$ or $\hat{p} \sim N(0.50, 0.00198)$

(b) $\hat{p} \sim N(0.90, \frac{(0.90)(0.10)}{70})$ or $\hat{p} \sim N(0.90, 0.001286)$

(c) $\hat{p} \sim N(0.63, \frac{(0.63)(0.37)}{100})$ or $\hat{p} \sim N(0.63, 0.002331)$

(d) $\hat{p} \sim N(0.26, \frac{(0.26)(0.74)}{243})$ or $\hat{p} \sim N(0.26, 0.000792)$

(8) $\hat{p} \sim N(0.29, \frac{(0.29)(0.71)}{72})$ or $\hat{p} \sim N(0.29, 0.00286)$

(9) $\hat{p} \sim N(0.19, \frac{(0.19)(0.81)}{100})$ or $\hat{p} \sim N(0.19, 0.001539)$

(10) $\hat{p} \sim N(0.02, \frac{(0.02)(0.98)}{860})$ or $\hat{p} \sim N(0.02, 0.00002279)$

Exercise 5.4

(1) 0.3520	**(2)** 0.9476	**(3)** 0.1210	**(4)** 0.2877	**(5)** 0.9161	**(6)** 0.0102
(7) 0.0162	**(8)** 0.0778	**(9)** 0.9568	**(10)** 0.3665		

Chapter 6

Exercise 6.1

(1) (a) $53.46 \leq \mu \leq 59.14$ (b) $53.62 \leq \mu \leq 58.97$ (c) $53.15 \leq \mu \leq 59.45$
(d) $52 \leq \mu \leq 60.6$ or $51.98 \leq \mu \leq 60.62$ (e) $53.37 \leq \mu \leq 59.23$

(2) (a) $31.3 \leq \mu \leq 35.78$ or $31.29 \leq \mu \leq 35.79$ (b) $31.51 \leq \mu \leq 35.57$
(c) $31.75 \leq \mu \leq 35.33$ (d) $32.29 \leq \mu \leq 34.79$ (e) $32.35 \leq \mu \leq 34.73$

(3) $4{,}504.59 \leq \mu \leq 4{,}621.41$ (4) $30.27 \leq \mu \leq 32.93$ (5) $2.406 \leq \mu \leq 2.594$
(6) $41.93 \leq \mu \leq 48.08$ (7) $73.38 \leq \mu \leq 82.16$ or $73.36 \leq \mu \leq 82.18$
(8) $7.51 \leq \mu \leq 8.09$ (9) $324.69 \leq \mu \leq 325.01$
(10) $1.312 \leq \mu \leq 1.388$

Exercise 6.2

(1) (a) $\mu \geq 34.42$ (b) $\mu \geq 34.33$ (c) $\mu \geq 34.44$ (d) $\mu \geq 33.94$ (e) $\mu \geq 34.25$
(f) $\mu \geq 34.39$

(2) (a) $\mu \leq 68.04$ (b) $\mu \leq 68.19$ (c) $\mu \leq 68.36$ (d) $\mu \leq 68.49$ or $\mu \leq 68.50$
(e) $\mu \leq 68.69$ (f) $\mu \leq 68.98$

(3) $\mu \geq 4{,}514.13$ or $\mu \geq 4{,}513.83$ (4) $\mu \leq 32.75$ (5) $\mu \leq 2.61$ (6) $\mu \geq 43.15$
(7) $\mu \leq 80.42$ or $\mu \leq 80.44$ (8) $\mu \geq 7.66$ or $\mu \geq 7.65$
(9) $\mu \leq 325.01$ (10) $\mu \geq 1.322$

Exercise 6.3

(1) $0.42 \leq \mu_x - \mu_y \leq 1.58$ (2) $-0.16 \leq \mu_x - \mu_y \leq 0.50$ (3) $0.63 \leq \mu_x - \mu_y \leq 1.63$
(4) $11.61 \leq \mu_x - \mu_y \leq 15.79$ (5) $1.31 \leq \mu_x - \mu_y \leq 4.69$ or $1.30 \leq \mu_x - \mu_y \leq 4.70$
(6) $54.96 \leq \mu_x - \mu_y \leq 60.24$ (7) $0.361 \leq \mu_x - \mu_y \leq 0.519$ (8) $89.88 \leq \mu_x - \mu_y \leq 101.52$
(9) $-0.003 \leq \mu_x - \mu_y \leq 0.025$ (10) $5.10 \leq \mu_x - \mu_y \leq 5.36$

Exercise 6.4

(1) (a) $0.669 \leq p \leq 0.851$ **(b)** $0.674 \leq p \leq 0.846$ **(c)** $0.681 \leq p \leq 0.839$
     (d) $0.653 \leq p \leq 0.867$ **(e)** $0.658 \leq p \leq 0.862$

(2) (a) $0.083 \leq p \leq 0.337$ **(b)** $0.104 \leq p \leq 0.316$ **(c)** $0.075 \leq p \leq 0.345$
     (d) $0.127 \leq p \leq 0.293$ **(e)** $0.05 \leq p \leq 0.37$

(3) $0.004 \leq p \leq 0.022$ **(4)** $0.199 \leq p \leq 0.321$ **(5)** $0.20 \leq p \leq 0.34$
(6) $0.112 \leq p \leq 0.308$ **(7)** $0.255 \leq p \leq 0.465$ **(8)** $0.015 \leq p \leq 0.125$
(9) $0.09 \leq p \leq 0.29$ **(10)** $0.122 \leq p \leq 0.238$

Exercise 6.5

(1) (a) $p \leq 0.834$ **(b)** $p \leq 0.827$ **(c)** $p \leq 0.819$ **(d)** $p \leq 0.851$ **(e)** $p \leq 0.846$

(2) (a) $p \geq 0.101$ **(b)** $p \geq 0.127$ **(c)** $p \geq 0.093$ **(d)** $p \geq 0.153$ **(e)** $p \geq 0.065$

(3) $p \geq 0.005$ **(4)** $p \leq 0.311$ **(5)** $p \leq 0.325$ **(6)** $p \geq 0.126$ **(7)** $p \leq 0.445$

(8) $p \geq 0.022$ **(9)** $p \leq 0.281$ **(10)** $p \geq 0.132$

Exercise 6.6

(1) $-0.26 \leq p_x - p_y \leq 0.063$ **(2)** $0.242 \leq p_x - p_y \leq 0.418$ **(3)** $0.636 \leq p_x - p_y \leq 0.804$

(4) $0.078 \leq p_x - p_y \leq 0.382$ **(5)** $0.038 \leq p_x - p_y \leq 0.202$ **(6)** $0.005 \leq p_x - p_y \leq 0.035$

(7) $-0.018 \leq p_x - p_y \leq 0.078$ **(8)** $-0.06 \leq p_x - p_y \leq 0.20$ **(9)** $0.55 \leq p_x - p_y \leq 0.77$

(10) $-0.10 \leq p_x - p_y \leq 0.26$

Chapter 7

Exercise 7.1

(a) Critical Values = ± 1.96; Test Statistic = - 2.81; p = 0.0025; $\frac{\alpha}{2}$ = 0.025; Reject *Ho*

(b) Critical Values = ± 1.96; Test Statistic = - 0.86; p = 0.1949; $\frac{\alpha}{2}$ = 0.025; Do Not Reject *Ho*

(c) Critical Values = ± 1.96; Test Statistic = - 6.31; p ≈ 0.0000; $\frac{\alpha}{2}$ = 0.025; Reject *Ho*

(d) Critical Values = ± 1.96; Test Statistic = 2.23; p = 0.0129; $\frac{\alpha}{2}$ = 0.025; Reject *Ho*

(e) Critical Values = ± 1.96; Test Statistic = 4.48; p ≈ 0.0000; $\frac{\alpha}{2}$ = 0.025; Reject *Ho*

(f) Critical Values = ± 1.96; Test Statistic = 1.30; p = 0.0968; $\frac{\alpha}{2}$ = 0.025; Do Not Reject *Ho*

Exercise 7.2

(1) Critical Values = ± 2.05; Test Statistic = - 2.32; p = 0.0102; $\frac{\alpha}{2}$ = 0.02; Reject *Ho*

(2) Critical Values = ± 2.17; Test Statistic = - 15.2; p ≈ 0.0000; $\frac{\alpha}{2}$ = 0.015; Reject *Ho*

(3) Critical Values = ± 1.81; Test Statistic = 1.81; p = 0.0351; $\frac{\alpha}{2}$ = 0.035; Do Not Reject *Ho*

(4) Critical Values = ± 1.88; Test Statistic = - 1.76; p = 0.0392; $\frac{\alpha}{2}$ = 0.03; Do Not Reject *Ho*

(5) Critical Values = ± 2.33; Test Statistic = 2.72; p = 0.0033; $\frac{\alpha}{2}$ = 0.01; Reject *Ho*

(6) Critical Values = ± 1.64/1.65; Test Statistic = 1.94; p = 0.0262; $\frac{\alpha}{2}$ = 0.05; Reject *Ho*

(7) Critical Values = ± 1.81; Test Statistic = -1.79; p = 0.0367; $\frac{\alpha}{2}$ = 0.035; Do Not Reject *Ho*

(8) Critical Values = ± 1.75; Test Statistic = - 1.68; p = 0.0465; $\frac{\alpha}{2}$ = 0.04; Do Not Reject *Ho*

(9) Critical Values = ± 1.70; Test Statistic = - 3.61; p ≈ 0.0000; $\frac{\alpha}{2}$ = 0.045; Reject *Ho*

(10) Critical Values = ± 1.96; Test Statistic = -10 ; p ≈ 0.0000; $\frac{\alpha}{2}$ = 0.025; Reject *Ho*

Exercise 7.3

(a) Critical Value = - 1.64; Test Statistic = - 2.81; p = 0.0025; α = 0.05; Reject Ho

(b) Critical Value = - 1.64; Test Statistic = - 0.86; p = 0.1949; α = 0.05; Do Not Reject Ho

(c) Critical Value = - 1.64; Test Statistic = - 6.31; p $\approx$ 0.0000; α = 0.05; Reject Ho

(d) Critical Value = -1.64; Test Statistic = - 4.09; p $\approx$ 0.0000; α = 0.05; Reject Ho

(e) Critical Value = -1.64; Test Statistic = - 2.20; p = 0.0139; α = 0.05; Reject Ho

(f) Critical Value = -1.64; Test Statistic = -1.22; p = 0.1112; α = 0.05; Do Not Reject Ho

Exercise 7.4

(a) Critical Values = 1.64; Test Statistic = 3.52; p $\approx$ 0.0000; α = 0.05; Reject Ho

(b) Critical Values = 1.64; Test Statistic = 0.85; p = 0.1977; α = 0.05; Do Not Reject Ho

(c) Critical Values = 1.64; Test Statistic = 1.28; p = 0.1003; α = 0.05; Do Not Reject Ho

(d) Critical Values = 1.64; Test Statistic = 1.64; p = 0.0505; α = 0.05; Do Not Reject Ho

(e) Critical Values = 1.64; Test Statistic = 7.99; p $\approx$ 0.0000; α = 0.05; Reject Ho

(f) Critical Values = 1.64; Test Statistic = 2.11; p = 0.0174; α = 0.05; Reject Ho

Exercise 7.5

(1) Critical Value = -1.88; Test Statistic = - 1.77; p = 0.0384; α = 0.03; Do Not Reject Ho

(2) Critical Value = 1.75; Test Statistic = 2.88; p = 0.0020; α = 0.02; Reject Ho

(3) Critical Value = -2.33; Test Statistic = - 5.51; p $\approx$ 0.0000; α = 0.05; Reject Ho

(4) Critical Value = 1.34; Test Statistic = 1.24 p = 0.1075; α = 0.09; Do Not Reject Ho

(5) Critical Value = -1.75; Test Statistic = -1.68; p = 0.0465; α = 0.04; Do Not Reject Ho

(6) Critical Value = -1.88; Test Statistic = -9; p ≈ 0.0000; α = 0.03; Reject *Ho*

(7) Critical Value = 2.33; Test Statistic = 2.82; p = 0.0024; α = 0.01; Reject *Ho*

(8) Critical Value = 2.05; Test Statistic = 3.48; p ≈ 0.0000; α = 0.02; Reject *Ho*

(9) Critical Value = -2.33; Test Statistic = -2.65; p = 0.0040; α = 0.01; Reject *Ho*

(10) (a) Critical Value = 1.75; Test Statistic = 1.81; p = 0.0351; α = 0.04; Reject *Ho*
     (b) Critical Value = 1.88; Test Statistic = 1.81; p = 0.0351; α = 0.03; Do Not Reject *Ho*

Exercise 7.6

(1) Critical Value = 2.05; Test Statistic = - 8.75; p ≈ 0.0000; α = 0.02; Reject *Ho*

(2) Critical Values = ± 2.17; Test Statistic = 1.71; p = 0.0436; α = 0.04; Do Not Reject *Ho*

(3) Critical Values = ± 2.57/2.58; Test Statistic = 3.38; p ≈ 0.0000; $\frac{\alpha}{2}$ = 0.005; Reject *Ho*

(4) Critical Values = 2.57/2.58; Test Statistic = -2.86; p = 0.0021; α = 0.005; Reject *Ho*

(5) Critical Values = 1.88; Test Statistic = 14.5; p ≈ 0.0000; α = 0.03; Reject *Ho*

(6) Critical Values = - 1.88; Test Statistic = - 2.29; p ≈ 0.0000; α = 0.03; Reject *Ho*

(7) (a) Critical Values = ± 2.33; Test Statistic = 2.51; p = 0.0060; $\frac{\alpha}{2}$ = 0.01; Reject *Ho*

    (b) Critical Values = ± 2.57/2.58; Test Statistic = 2.51; p = 0.0060; $\frac{\alpha}{2}$ = 0.005;

      Do Not Reject *Ho*

(8) (a) Critical Values = ± 1.96; Test Statistic = - 4.05; p ≈ 0.0000 ; $\frac{\alpha}{2}$ = 0.025; Reject *Ho*

    (b) Critical Values = ± 1.96; Test Statistic = 4.05; p ≈ 0.0000; $\frac{\alpha}{2}$ = 0.025; Reject *Ho*

      Both tests are the same.

(9) Critical Values = ± 2.17; Test Statistic = 2.07; p = 0.0192; $\frac{\alpha}{2}$ = 0.015; Do Not Reject H_0

(10) Critical Values = ± 2.57/ ± 2.58; Test Statistic = 3.21; p ≈ 0.0000 ; $\frac{\alpha}{2}$ = 0.005;

    Reject *Ho*

Exercise 7.7

(a) Critical Values = ± 1.96; Test Statistic = - 1.31; p = 0.0951; $\frac{\alpha}{2}$ = 0.025; Do Not Reject H_O

(b) Critical Values = ± 1.96; Test Statistic = 1.17; p = 0.1210; $\frac{\alpha}{2}$ = 0.025; Do Not Reject H_O

(c) Critical Values = ± 1.96; Test Statistic = - 0.55; p = 0.2912; $\frac{\alpha}{2}$ = 0.025; Do Not Reject H_O

(d) Critical Values = ± 1.96; Test Statistic = 0.52; p = 0.3015; $\frac{\alpha}{2}$ = 0.025; Do Not Reject H_O

(e) Critical Values = ± 1.96; Test Statistic = - 2.48; p = 0.0066; $\frac{\alpha}{2}$ = 0.025; Reject H_O

(f) Critical Values = ± 1.96; Test Statistic = 2.77; p = 0.0028; $\frac{\alpha}{2}$ = 0.025; Do Not Reject H_O

Exercise 7.8

(1) Critical Values = ± 1.28; Test Statistic = 0.91; p = 0.1814; $\frac{\alpha}{2}$ = 0.10; Do Not Reject H_O

(2) (a) Critical Values = ± 2.17; Test Statistic = -2.35; p = 0.0094; $\frac{\alpha}{2}$ = 0.015; Reject H_O

    (b) Critical Values = ± 2.57/2.58; Test Statistic = - 2.35; p = 0.0094; $\frac{\alpha}{2}$ = 0.005; Do Not
       Reject H_O

(3) Critical Values = ± 1.81; Test Statistic = - 1.86; p = 0.0314; $\frac{\alpha}{2}$ = 0.035; Reject H_O

(4) Critical Values = ± 2.05; Test Statistic = 2.07; p = 0.0192; $\frac{\alpha}{2}$ = 0.02; Reject H_O

(5) Critical Values = ± 1.75; Test Statistic = 1. 48; p = 0.0694; $\frac{\alpha}{2}$ = 0.04; Do Not Reject H_O

(6) Critical Values = ± 2.33; Test Statistic = 2.25; p = 0.0122; $\frac{\alpha}{2}$ = 0.01; Do Not Reject H_O

(7) Critical Values = ± 1.96; Test Statistic = - 0.77; p = 0.2206; $\frac{\alpha}{2}$ = 0.025; Do Not Reject H_O

(8) Critical Values = ± 1.88; Test Statistic = -1.70; p = 0.0446; $\frac{\alpha}{2}$ = 0.03; Do Not Reject H_O

(9) Critical Values = ± 2.17; Test Statistic = -2.29; p = 0.0110; $\frac{\alpha}{2}$ = 0.005; Reject H_O

(10) Critical Values = ± 1.60; Test Statistic = - 0.84; p = 0.2005; $\frac{\alpha}{2}$ = 0.055; Do Not Reject H_O

Exercise 7.9

(a) Critical Value = - 1.64; Test Statistic = - 0.88; p = 0.0375; α = 0.05; Reject *Ho*

(b) Critical Value = - 1.64; Test Statistic = - 0.29; p $\approx$ 0.0000; α = 0.05; Reject *Ho*

(c) Critical Value = - 1.64; Test Statistic = - 0.55; p = 0.2981; α = 0.05; Do Not Reject *Ho*

(d) Critical Value = - 1.64; Test Statistic = - 2.23; p = 0.0244; α = 0.05; Reject *Ho*

(e) Critical Value = - 1.64; Test Statistic = -1.46; p = 0.0918; α = 0.05; Do Not Reject *Ho*

(f) Critical Value = -1.64; Test Statistic = -1.64; p = 0.0505; α = 0.05; Do Not Reject *Ho*

Exercise 7.10

(a) Critical Value = 1.64; Test Statistic = 0.58; p = 0.2810; α = 0.05; Do Not Reject *Ho*

(b) Critical Value = 1.64; Test Statistic = 0.39; p = 0.3483; α = 0.05; Do Not Reject *Ho*

(c) Critical Value = 1.64; Test Statistic = 0.91; p = 0.1814; α = 0.05; Do Not Reject *Ho*

(d) Critical Value = 1.64; Test Statistic = 2.19; p = 0.0143; α = 0.05; Reject *Ho*

(e) Critical Value = 1.64; Test Statistic = 1.17; p = 0.1210; α = 0.05; Do Not Reject *Ho*

(f) Critical Value = 1.64; Test Statistic = 1.66; p = 0.0485; α = 0.05; Reject *Ho*

Exercise 7.11

(1) Critical Value = 1.75; Test Statistic = 1.82; p = 0.0344; α = 0.04; Reject *Ho*

(2) Critical Value = - 2.33; Test Statistic = -3.26; p $\approx$ 0.0000; α = 0.01; Reject *Ho*

(3) Critical Value = 2.05; Test Statistic = 3.46; p $\approx$ 0.0000; α = 0.02; Reject *Ho*

(4) Critical Value = 1.41; Test Statistic = 1.36; p = 0.0869; α = 0.08; Do Not Reject *Ho*

(5) Critical Value = -1.64/-1.65; Test Statistic = -2.68; p = 0.0037; α = 0.05; Reject *Ho*

(6) (a) Critical Value = -2.05; Test Statistic = -2.22; p = 0.0132; $\alpha = 0.02$; Reject *Ho*
    (b) Critical Value = -2.33; Test Statistic = -2.22; p = 0.0132; $\alpha = 0.01$; Do Not Reject *Ho*

(7) (a) Critical Value = 1.75; Test Statistic = 1.88; p = 0.0301; $\alpha = 0.04$; Reject *Ho*
    (b) Critical Value = 1.88 Test Statistic = 1.88; p = 0.0301; $\alpha = 0.03$; Do Not Reject *Ho*

(8) Critical Value = -1.34; Test Statistic = -1.19; p = 0.1170; $\alpha = 0.09$; Do Not Reject *Ho*

(9) Critical Values = 2.33; Test Statistic = 2.77; p = 0.0028; $\alpha = 0.01$; Reject *Ho*

(10) Critical Values = 2.05; Test Statistic = 1.99; p = 0.0233; $\alpha = 0.02$; Reject *Ho*

Exercise 7.12
(1) Critical Value = 2.33; Test Statistic = 2.37; p = 0.0089; $\alpha = 0.01$; Reject *Ho*

(2) Critical Value = -1.28; Test Statistic = - 0.14; p = 0. 4443; $\alpha = 0.10$; Do Not Reject *Ho*

(3) Critical Values = ±1.88; Test Statistic = 1.07; p = 0.1423; $\frac{\alpha}{2} = 0.03$; Do Not Reject *Ho*

(4) Critical Values = ±1.28; Test Statistic = 1.05; p = 0.1469; $\frac{\alpha}{2} = 0.015$; Do Not Reject *Ho*

(5) Critical Value = 1.34; Test Statistic = 0.51; p = 0.3050; $\alpha = 0.09$; Do Not Reject *Ho*

(6) Critical Value = -1.88; Test Statistic = -2.01; p = 0.0222; $\alpha = 0.03$; Reject *Ho*

(7) Critical Value = 2.33; Test Statistic = 4.05; p $\approx$ 0.0000; $\alpha = 0.01$; Reject *Ho*

(8) Critical Value = - 2.05; Test Statistic = -1.20; p = 0.1151; $\alpha = 0.02$; Do Not Reject *Ho*

(9) (a) Critical Values = ±2.57/2.58; Test Statistic = 2.14; p = 0.0162; $\frac{\alpha}{2} = 0.005$; Do Not
    Reject *Ho*

    (b) Critical Values = ±2.17; Test Statistic = 2.14; p = 0.0162; $\frac{\alpha}{2} = 0.015$; Do Not Reject *Ho*

    (c) Critical Values = ±1.96; Test Statistic =2.14; p = 0.0162; $\frac{\alpha}{2} = 0.025$; Reject *Ho*

    (d) Critical Values = ±1.81; Test Statistic = 2.14; p = 0.0162; $\frac{\alpha}{2} = 0.035$; Reject *Ho*

(10) Critical Value = ±1.96; Test Statistic = 0.99; p = 0.1611; $\frac{\alpha}{2} = 0.025$; Do Not Reject *Ho*

Appendix C
Index

www.ingramcontent.com/pod-product-compliance
Lightning Source LLC
Chambersburg PA
CBHW041705210326
41598CB00007B/545